Praise for
DAVID EAGLEMAN'S

THE BRAIN

"Eagleman wants to make you more conscious. . . . For him, neuroscience is an inroad to self-knowledge, which is itself an inroad to solutions for social problems as wide-ranging as how to reform criminals and preventing genocide."

—*New York* Magazine

"Stimulating. . . . He effectively unveils the stunning degree to which 'we can now hack our own hardware' in order to understand, and better, ourselves."

—*Publishers Weekly*

"David Eagleman's *The Brain* is an astonishing read. On every page there is a revelation so fantastic as to make one gasp. It would be impossible to take in if we didn't all possess that impossibly extraordinary thing, a brain. Eagleman comes closer than anyone to solving the mystery of how to find the self inside the grey electric mush between our ears."

—Stephen Fry, author of *The Ode Less Travelled*

"Eagleman is . . . the hot neuroscientist at a time when the brain is the hot organ."

—*Houston Chronicle*

"Mind-blowing revelations abound."

—*Financial Times*

THE BRAIN

DAVID EAGLEMAN

David Eagleman is a neuroscientist at Stanford University. His scientific research is published in journals from *Science* to *Nature*, and he is also the author of the internationally bestselling books *Sum* and *Incognito*. He is the writer and presenter of the companion PBS television series *The Brain*.

www.eagleman.com

Also by David Eagleman

Sum: Forty Tales from the Afterlives

Incognito: The Secret Lives of the Brain

Why the Net Matters: Six Easy Ways to Avert the Collapse of Civilization

With Richard Cytowic

Wednesday is Indigo Blue: Discovering the Brain of Synesthesia

THE BRAIN

THE STORY OF YOU

DAVID EAGLEMAN

VINTAGE BOOKS
A Division of Penguin Random House LLC
New York

FIRST VINTAGE BOOKS EDITION, MARCH 2017

 Published in the United States by Vintage Books, a division of Penguin Random House LLC, New York, and distributed in Canada by Random House of Canada, a division of Penguin Random House Canada Ltd., Toronto. Originally published in hardcover in the United States by Pantheon Books, a division of Penguin Random House LLC, New York, in 2015. Simultaneously published in Great Britain by Canongate Books Ltd., Edinburgh.

The Cataloging-in-Publication Data is available from the Library of Congress.

Vintage Trade Paperback ISBN: 978-0-525-43344-6
eBook ISBN: 978-1-101-87054-9

www.vintagebooks.com

Printed in the United States of America
10 9 8 7 6 5 4 3 2

Contents

THE BRAIN

THE STORY OF YOU

INTRODUCTION

Because brain science is a fast-moving field, it's rare to step back to view the lay of the land, to work out what our studies mean for our lives, to discuss in a plain and simple way what it means to be a biological creature. This book sets out to do that.

Brain science matters. The strange computational material in our skulls is the perceptual machinery by which we navigate the world, the stuff from which decisions arise, the material from which imagination is forged. Our dreams and our waking lives emerge from its billions of zapping cells. A better understanding of the brain sheds light on what we take to be real in our personal relationships and what we take to be necessary in our social policy: how we fight, why we love, what we accept as true, how we should educate, how we can craft better social policy, and how to design our bodies for the centuries to come. In the brain's microscopically small circuitry is etched the history and future of our species.

Given the brain's centrality to our lives, I used to wonder why our society so rarely talks about it, preferring instead to fill our airwaves with celebrity gossip and reality shows. But I now think this lack of attention to the brain can be taken not as a shortcoming, but as a clue: we're so trapped inside our reality that it is inordinately difficult to realize we're trapped inside anything. At first blush, it seems that perhaps there's nothing to talk about. Of course colors exist in the outside world. Of course my memory is like a video camera. Of course I know the real reasons for my beliefs.

The pages of this book will put all our assumptions under the spotlight. In writing it, I wanted to get away from a textbook model in favor of illuminating a deeper level of inquiry: how we decide, how we perceive reality, who we are, how our lives are steered, why we need other people, and where we're heading as a species that's just beginning to grab its own reins. This project attempts to bridge the gap between the academic literature and the lives we lead as brain owners. The approach I take here diverges from the academic journal articles I write, and even from my other neuroscience books. This project is meant for a different kind of audience. It doesn't presuppose any specialized knowledge, only curiosity and an appetite for self-exploration.

So strap in for a whistle-stop tour into the inner cosmos. In the infinitely dense tangle of billions of brain cells and their trillions of connections, I hope you'll be able to squint and make out something that you might not have expected to see in there. You.

1

WHO AM I?

All the experiences in your life—from single conversations to your broader culture—shape the microscopic details of your brain. Neurally speaking, who you are depends on where you've been. Your brain is a relentless shape-shifter, constantly rewriting its own circuitry—and because your experiences are unique, so are the vast, detailed patterns in your neural networks. Because they continue to change your whole life, your identity is a moving target; it never reaches an endpoint.

Although neuroscience is my daily routine, I'm still in awe every time I hold a human brain. After you take into account its substantial weight (an adult brain weighs in at three pounds), its strange consistency (like firm jelly), and its wrinkled appearance (deep valleys carving a puffy landscape)—what's striking is the brain's sheer physicality: this hunk of unremarkable stuff seems so at odds with the mental processes it creates.

Our thoughts and our dreams, our memories and experiences all arise from this strange neural material. Who we are is found within its intricate firing patterns of electrochemical pulses. When that activity stops, so do you. When that activity changes character, due to injury or drugs, you change character in lockstep. Unlike any other part of your body, if you damage a small piece of the brain, who you are is likely to change radically. To understand how this is possible, let's start at the beginning.

BORN UNFINISHED

At birth we humans are helpless. We spend about a year unable to walk, about two more before we can articulate full thoughts, and many more years unable to fend for ourselves. We are totally dependent on those around us for our survival. Now compare this to many other mammals. Dolphins, for instance, are born swimming; giraffes learn to stand within hours; a baby zebra can run within forty-five minutes of birth. Across the animal kingdom, our cousins are strikingly independent soon after they're born.

On the face of it, that seems like a great advantage for other species—but in fact it signifies a limitation. Baby animals develop quickly because their brains are wiring up according to a largely preprogrammed routine. But that preparedness trades off with flexibility. Imagine if some hapless rhinoceros found itself on the Arctic tundra, or on top of a mountain in the Himalayas, or in the middle of urban Tokyo. It would have no capacity to adapt (which is why we don't find rhinos in those areas). This strategy of arriving with a prearranged brain works inside a particular niche in the ecosystem—but put an animal outside of that niche, and its chances of thriving are low.

In contrast, humans are able to thrive in many different environments, from the frozen tundra to the high mountains to bustling urban centers. This is possible because the human brain is born remarkably unfinished. Instead of arriving with everything wired up—let's call it "hardwired"—a human brain allows itself to be shaped by the details of life experience. This leads to long periods of helplessness as the young brain slowly molds to its environment. It's "livewired."

CHILDHOOD PRUNING: EXPOSING THE STATUE IN THE MARBLE

What's the secret behind the flexibility of young brains? It's not about growing new cells—in fact, the number of brain cells is the same in children and adults. Instead, the secret lies in how those cells are connected.

At birth, a baby's neurons are disparate and unconnected, and in the first two years of life they begin connecting up extremely rapidly as they take in sensory information. As many as two million new connections, or synapses, are formed every second in an infant's brain. By age two, a child has over one hundred trillion synapses, double the number an adult has.

It has now reached a peak and has far more connections than it will need. At this point, the blooming of new connections is supplanted by a strategy of neural "pruning." As you mature, 50 percent of your synapses will be pared back.

Which synapses stay and which go? When a synapse successfully participates in a circuit, it is strengthened; in contrast, synapses weaken if they aren't useful, and eventually they are eliminated. Just like paths in a forest, you lose the connections that you don't use.

In a sense, the process of becoming who you are is defined by carving back the possibilities that were already present. You become who you are not because of what grows in your brain, but because of what is removed.

LIVEWIRING

Many animals are born genetically preprogrammed, or "hardwired" for certain instincts and behaviors. Genes guide the construction of their bodies and brains in specific ways that define what they will be and how they'll behave. A fly's reflex to escape in the presence of a passing shadow; a robin's preprogrammed instinct to fly south in the winter; a bear's desire to hibernate; a dog's drive to protect its master: these are all examples of instincts and behaviors that are hardwired. Hardwiring allows these creatures to move as their parents do from birth, and in some cases to eat for themselves and survive independently.

In humans the situation is somewhat different. The human brain comes into the world with some amount of genetic hardwiring (for example, for breathing, crying, suckling, caring about faces, and having the ability to learn the details of their native language). But compared to the rest of the animal kingdom, human brains are unusually incomplete at birth. The detailed wiring diagram of the human brain is not preprogrammed; instead, genes give very general directions for the blueprints of neural networks, and world experience fine-tunes the rest of the wiring, allowing it to adapt to the local details.

The human brain's ability to shape itself to the world into which it's born has allowed our species to take over every ecosystem on the planet and begin our move into the solar system.

Throughout our childhoods, our local environments refine our brain, taking the jungle of possibilities and shaping it back to correspond to what we're exposed to. Our brains form fewer but stronger connections.

As an example, the language that you're exposed to in infancy (say, English versus Japanese) refines your ability to hear the particular sounds of your language, and worsens your capacity to hear the sounds of other languages. That is, a baby born in Japan and a baby born in America can hear and respond to all the sounds in both languages. Through time, the baby raised in Japan will lose the ability to distinguish between, say, the sounds of R and L, two sounds that aren't separated in Japanese. We are sculpted by the world we happen to drop into.

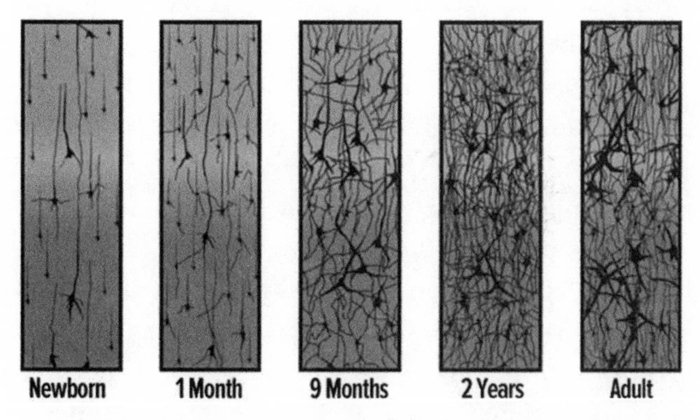

In a newborn brain, neurons are relatively unconnected to one another. Over the first two to three years, the branches grow and the cells become increasingly connected. After that, the connections are pruned back, becoming fewer and stronger in adulthood.

NATURE'S GAMBLE

Over our protracted childhood, the brain continually pares back its connections, shaping itself to the particulars of its environment. This is a smart strategy to match a brain to its environment—but it also comes with risks.

If developing brains are not given the proper, "expected" environment—one in which a child is nurtured and looked after—the brain will struggle to develop normally. This is something the Jensen family from Wisconsin has experienced firsthand. Carol and Bill Jensen adopted Tom, John, and Victoria when the children were four years old. The three children were orphans who had, until their adoption, endured appalling conditions in state-run orphanages in Romania—with consequences for their brain development.

When the Jensens picked up the children and took a taxi out of Romania, Carol asked the taxi driver to translate what the children were saying. The taxi driver explained they were speaking gibberish. It was not a known language; starved of normal interaction, the children had developed a strange creole. As they've grown up, the children have had to deal with learning disabilities, the scars of their childhood deprivation.

Tom, John, and Victoria don't remember much about their time in Romania. In contrast, someone who remembers the institutions vividly is Dr. Charles Nelson, Professor of Pediatrics at Boston Children's Hospital. He first visited these institutions in 1999.

What he saw horrified him. Young children were kept in their cribs, with no sensory stimulation. There was a single caretaker for every fifteen children, and these workers were instructed not to pick the children up or show them affection in any way, even when they were crying—the concern was that such displays of affection would lead to the children wanting more, an impossibility with the limited staffing. In this context, things were as regimented as possible. Children were lined up on plastic pots for toileting. Everyone got the same haircut, regardless of gender. They were dressed alike, fed on schedule. Everything was mechanized.

Children whose cries went unanswered soon learned not to cry. The children were not held and were not played with. Although they had their basic needs met (they were fed, washed and clothed), the infants were deprived of emotional care, support, and any kind of stimulation. As a result, they developed "indiscriminate friendliness." Nelson explains that he'd walk into a room and be surrounded by little kids he'd never seen before—and they'd want to jump into his arms and sit on his lap or hold his hand or walk off with him. Although this sort of indiscriminate behavior seems sweet at first glance, it's a coping strategy of neglected children, and it goes hand-in-hand with long-term attachment issues. It is a hallmark behavior of children who have grown up in an institution.

Shaken by the conditions they were witnessing, Nelson and his team set up the Bucharest Early

ROMANIA'S ORPHANAGES

In 1966, to increase the population and the workforce, Romanian president Nicolae Ceauşescu banned contraception and abortion. State gynecologists known as "menstrual police" examined women of childbearing age to ensure they were producing enough offspring. A "celibacy tax" was levied on families who had fewer than five children. The birth rate skyrocketed.

Many poor families couldn't afford to care for their children—and so they gave them over to state-run institutions. In turn, the state rolled out more institutions to meet the soaring numbers. By 1989, when Ceauşescu was ousted, 170,000 abandoned children resided in institutions.

Scientists soon revealed the consequences of an institutional upbringing on brain development. And those studies influenced government policy. Over the years, most of the Romanian orphans have been returned to their parents or removed to government foster care. By 2005, Romania made it illegal for children to be institutionalized before the age of two, unless severely disabled.

Millions of orphans still live in institutionalized government care around the world. Given the necessity of a nurturing environment for an infant's developing brain, it is imperative that governments find ways to get the children into conditions that allow proper brain development.

Intervention Program. They assessed 136 children, aged six months to three years, who had been living in institutions from birth. First, it became clear that the children had IQs in the sixties and seventies, compared to an average of one hundred. The children showed signs of underdeveloped brains and their language was very delayed. When Nelson used electroencephalography (EEG) to measure the electrical activity in these children's brains, he found they had dramatically reduced neural activity.

Without an environment with emotional care and cognitive stimulation, the human brain cannot develop normally.

Encouragingly, Nelson's study also revealed an important flip side: the brain can often recover, to varying degrees, once the children are removed to a safe and loving environment. The younger a child is removed, the better his recovery. Children removed to foster homes before the age of two generally recovered well. After two, they made improvements—but depending on the age of the child they were left with differing levels of developmental problems.

Nelson's results highlight the critical role of a loving, nurturing environment for a developing child's brain. And this illustrates the profound importance of the environment around us in shaping who we become. We are exquisitely sensitive to our surroundings. Because of the wire-on-the-fly strategy of the human brain, who we are depends heavily on where we've been.

THE TEEN YEARS

Only a couple of decades ago it was thought that brain development was mostly complete by the end of childhood. But we now know that the process of building a human brain takes up to twenty-five years. The teen years are a period of such important neural reorganization and change that it dramatically affects who we seem to be. Hormones coursing around our bodies cause obvious physical changes as we take on the appearance of adults—but out of sight our brains are undergoing equally monumental changes. These changes profoundly color how we behave and react to the world around us.

One of these changes has to do with an emerging sense of self—and with it, self-consciousness.

To get a sense of the teen brain at work, we ran a simple experiment. With the help of my graduate student Ricky Savjani, we asked volunteers to sit on a stool in a shop window display. We then pulled back the curtain to expose the volunteer looking out on the world—to be gawked at by passersby.

Before sending them into this socially awkward situation, we rigged up each volunteer so we'd be able to measure their emotional response. We hooked them up with a device to measure the galvanic skin response (GSR), a useful proxy for anxiety: the more your sweat glands open, the higher your skin conductance will be. (This is, by the way, the same technology used in a lie detector, or polygraph test.)

SCULPTING OF THE ADOLESCENT BRAIN

After childhood, just before the onset of puberty, there is a second period of overproduction: the prefrontal cortex sprouts new cells and new connections (synapses), thereby creating new pathways for molding. This excess is followed by approximately a decade of pruning: all through our teenage years, weaker connections are trimmed back while stronger connections are reinforced. As a result of this thinning out, the volume of the prefrontal cortex reduces by about 1 percent per year during the teenage years. The shaping of circuits during the teen years sets us up for the lessons we learn on our paths to becoming adults.

Because these massive changes take place in brain areas required for higher reasoning and the control of urges, adolescence is a time of steep cognitive change. The dorsolateral prefrontal cortex, important for controlling impulses, is among the most belated regions to mature, not reaching its adult state until the early twenties. Well before neuroscientists worked out the details, car insurance companies noticed the consequences of incomplete brain maturation—and they accordingly charge more for teen drivers. Likewise, the criminal justice system has long held this intuition, and thus juveniles are treated differently than adults.

Both adults and teens participated in our experiment. In adults, we observed a stress response from being stared at by strangers, exactly as expected. But in teenagers, that same experience caused social emotions to go into overdrive: the teens were much more anxious—some to the point of trembling—while they were being watched.

Why the difference between the adults and teens? The answer involves an area of the brain called the medial prefrontal cortex (mPFC). This region becomes active when you think about your self—and especially the emotional significance of a situation to your self. Dr. Leah Somerville and her colleagues at Harvard University found that as one grows from childhood to adolescence, the mPFC becomes more active in social situations, peaking at around fifteen years old. At this point, social situations carry a lot of emotional weight, resulting in a self-conscious stress response of high intensity. That is, in adolescence, thinking about one's self—so-called "self evaluation"—is a high priority. In contrast, an adult brain has grown accustomed to a sense of self—like having broken in a new pair of shoes—and as a result an adult doesn't care as much about sitting in the shop window.

Beyond social awkwardness and emotional hypersensitivity, the teen brain is set up to take risks. Whether it's driving fast or sexting naked photos, risky behaviors are more tempting to the teen brain than to the adult brain. Much of that has to do with the way we respond to rewards and incentives. As we move from childhood into adolescence, the brain shows an increasing response to rewards in areas related to pleasure seeking (one such area is called the nucleus accumbens). In teens, the activity here is as high as it is in adults. But here's the important fact: activity in the orbitofrontal cortex—involved in executive decision making, attention, and simulating future consequences—is still about the same in teens as it is in children. A mature pleasure-seeking system coupled with an immature orbitofrontal cortex means that teens are not only emotionally hypersensitive, but also less able to control their emotions than adults.

Moreover, Somerville and her team have an idea why peer pressure strongly compels behavior in teens: areas involved in social considerations (such as the mPFC) are more strongly coupled to other brain regions that translate motivations into actions (the striatum and its network of connections). This, they suggest, might explain why teens are more likely to take risks when their friends are around.

How we see the world as a teenager is the consequence of a changing brain that's right on schedule. These changes lead us to be more self-conscious, more risk-taking, and more prone to peer-motivated behavior. For frustrated parents the world over, there's an important message: who we are as a teenager is not simply the result of a choice or an attitude; it is the product of a period of intense and inevitable neural change.

PLASTICITY IN ADULTHOOD

By the time we're twenty-five years of age, the brain transformations of childhood and adolescence are finally over. The tectonic shifts in our identity and personality have ended, and our brain appears to now be fully developed. You might think that who we are as adults is now fixed in place, immovable. But it's not: in adulthood our brains continue to change. Something that can be shaped—and can hold that shape—is what we describe as plastic. And so it is with the brain, even in adulthood: experience changes it, and it retains the change.

To get a sense of how impressive these physical changes can be, consider the brains of a particular group of men and women who work in London: the city's cab drivers. They undergo four years of intensive training to pass the "Knowledge of London," one of society's most difficult feats of memory. The Knowledge requires aspiring cabbies to memorize London's extensive roadways, in all their combinations and permutations. This is an exceedingly difficult task: The Knowledge covers 320 different routes through the city, 25,000 individual streets, and 20,000 landmarks and points of interest—hotels, theaters, restaurants, embassies, police stations, sports facilities, and anywhere a passenger is likely to want to go. Students of The Knowledge typically spend three to four hours a day reciting theoretical journeys.

The unique mental challenges of The Knowledge

sparked the interest of a group of neuroscientists from University College London, who scanned the brains of several cab drivers. The scientists were particularly interested in a small area of the brain called the hippocampus—vital for memory, and, in particular, spatial memory.

The scientists discovered visible differences in the cabbies' brains: in the drivers, the posterior part of the hippocampus had grown physically larger than those in the control group—presumably causing their increased spatial memory. The researchers also found that the longer a cabbie has been doing his job, the bigger the change in that brain region, suggesting that the result was not simply reflecting a preexisting condition of people who go into the profession, but instead resulted from practice.

The cab-driver study demonstrates that adult brains are not fixed in place, but instead can reconfigure so much that the change is visible to the trained eye.

It's not just cab drivers whose brains reshape themselves. When one of the most famous brains of the twentieth century was examined, Albert Einstein's brain did not reveal the secret of his genius. But it did show that the brain area devoted to his left fingers had expanded—forming a giant fold in his cortex called the Omega sign, shaped like the Greek symbol Ω—all thanks to his less commonly known passion for playing the violin. This fold becomes enlarged in experienced violin players, who intensively develop fine dexterity with the fingers of their left hand. Piano players, in contrast, develop an Omega sign in both hemispheres, as they use both hands in fine, detailed movements.

The shape of the hills and valleys in the brain is largely conserved across people—but the finer details give a personal and unique reflection of where you've been and who you are now. Although most of the changes are too small to detect with the naked eye, everything you've experienced has altered the physical structure of your brain—from the expression of genes to the positions of molecules to the architecture of neurons. Your family of origin, your culture, your friends, your work, every movie you've watched, every conversation you've had—these have all left their footprints in your nervous system. These indelible, microscopic impressions accumulate to make you who you are, and to constrain who you can become.

PATHOLOGICAL CHANGES

Changes in our brain represent what we've done and who we are. But what happens if the brain changes because of a disease or injury? Does this also alter who we are, our personalities, our actions?

On August 1, 1966, Charles Whitman took an elevator to the observation deck of the University of Texas Tower in Austin. Then the twenty-five-year-old started firing indiscriminately at people below. Thirteen people were killed and thirty-three wounded, until Whitman himself was finally shot dead by police. When they got to his house they discovered that he had killed his wife and mother the night before.

There was only one thing more surprising than this random act of violence, and that was the lack of anything about Charles Whitman that would seem to have predicted it. He was an Eagle Scout, he was employed as a bank teller, and he was an engineering student.

Shortly after killing his wife and his mother, he'd sat down and typed what amounted to a suicide note:

> *I don't really understand myself these days. I am supposed to be an average reasonable and intelligent young man. However, lately (I cannot recall when it started) I have been a victim of many unusual and irrational thoughts . . . After my death I wish that an autopsy would be performed on me to see if there is any visible physical disorder.*

Whitman's request was granted. After an autopsy, the pathologist reported that Whitman had a small brain tumor. It was about the size of a nickel, and it was pressing against a part of his brain called the amygdala, which is involved in fear and aggression. This small amount of pressure on the amygdala led to a cascade of consequences in Whitman's brain, resulting in him taking actions that would otherwise be completely out of character. His brain matter had been changing, and who he was changed with it.

This is an extreme example, but less dramatic

changes in your brain can alter the fabric of who you are. Consider the ingestion of drugs or alcohol. Particular kinds of epilepsy make people more religious. Parkinson's disease often makes people lose their faith, while the medication for Parkinson's can often turn people into compulsive gamblers. It's not just illness or chemicals that change us: from the movies we watch to the jobs we work, everything contributes to a continual reshaping of the neural networks we summarize as us. So who exactly are you? Is there anyone down deep, at the core?

AM I THE SUM OF MY MEMORIES?

Our brains and bodies change so much during our life that—like a clock's hour hand—it's difficult to detect the changes. Every four months your red blood cells are entirely replaced, for instance, and your skin cells are replaced every few weeks. Within about seven years every atom in your body will be replaced by other atoms. Physically, you are constantly a new you. Fortunately, there may be one constant that links all these different versions of your self together: memory. Perhaps memory can serve as the thread that makes you who you are. It sits at the core of your identity, providing a single, continuous sense of self.

But there might be a problem here. Could the continuity be an illusion? Imagine you could walk into a park and meet your self at different ages in your life. There you are aged six; as a teenager; in your late twenties; midfifties; early seventies; all the way through your final years. In this scenario, you could all sit together and share the same stories about your life, teasing out the single thread of your identity.

Or could you? You all possess the name and history, but the fact is that you're all somewhat different people, in possession of different values and goals. And your life's memories might have less in common than expected. Your memory of who you were at fifteen is different from who you actually were at fifteen; moreover, you'll have different memories that relate back to the same events. Why? Because of what a memory is—and isn't.

Rather than memory being an accurate video recording of a moment in your life, it is a fragile brain state from a bygone time that must be resurrected for you to remember.

Here's an example: you're at a restaurant for a friend's birthday. Everything you experience triggers particular patterns of activity in your brain. For example, there's a particular pattern of activity sparked into life by the conversation between your friends. Another pattern is activated by the smell of the coffee; yet another by the taste of a delicious little French cake. The fact that the waiter puts his thumb in your cup is another memorable detail, represented by a different configuration of neurons firing. All of these constellations become linked with one another in a vast associative network of neurons that the hippocampus replays, over and over, until the associations become fixed. The neurons that are active at the same time will establish stronger connections between them: cells that fire together, wire together. The resulting network is the unique signature of the event, and it represents your memory of the birthday dinner.

Now let's imagine that six months later you taste one of those little French cakes, just like the one you tasted at the birthday party. This very specific key can unlock the whole web of associations. The original constellation lights up, like the lights of a city switching on. And suddenly you're back in that memory.

Although we don't always realize it, the memory is not as rich as you might have expected. You know that your friends were there. He must have been wearing a suit, because he always wears a suit. And she was wearing a blue shirt. Or maybe it was purple? It might have been green. If you really probe the memory, you'll realize that you can't remember the details of any of the other diners at the restaurant, even though the place was full.

So your memory of the birthday meal has started to fade. Why? For a start, you have a finite number of neurons, and they are all required to multitask. Each neuron participates in different constellations at different times. Your neurons operate in a dynamic matrix of shifting relationships, and heavy demand is continually placed on them to wire with others. So your memory of the birthday dinner has become muddied, as those "birthday" neurons have been co-opted to participate in other memory networks. The enemy of memory isn't time; it's other memories. Each new event needs to establish new relationships among a finite number of neurons. The surprise is that a faded memory doesn't seem faded to you. You feel, or at least assume, that the full picture is there.

And your memory of the event is even more dubious. Say that in the intervening year since the dinner, your two friends have split up. Thinking back on the dinner, you might now misremember sensing red flags. Wasn't he more quiet than usual that night? Weren't there moments of awkward silence between the two? Well, it will be difficult to know for certain, because the knowledge that's in your network now changes the memory that corresponds to then. You can't help but have your present color your past. So a single event may be perceived somewhat differently by you at different stages in your life.

THE FALLIBILITY OF MEMORY

Clues to the malleability of our memory come from the pioneering work of Professor Elizabeth Loftus at University of California, Irvine. She transformed the field of memory research by showing how susceptible memories are.

Loftus devised an experiment in which she invited volunteers to watch films of car crashes, and then asked them a series of questions to test what they remembered. The questions she asked influenced the answers she received. She explains: "When I asked how fast were the cars going when they hit each other, versus how fast were the cars going when they smashed into each other, witnesses give different estimates of speed. They thought the cars were going faster when I used the word smashed." Intrigued by the way that leading questions could contaminate memory, she decided to go further.

Would it be possible to implant entirely false memories? To find out, she recruited a selection of participants, and had her team contact their families to get information about events in their past. Armed with this information, the researchers put together four stories about each participant's childhood. Three were true. The fourth story contained plausible information, but was entirely made up. The fourth story was about getting lost in a shopping mall as a child, being found by a kind elderly person, and finally being reunited with a parent.

In a series of interviews, participants were told the four stories. At least a quarter claimed they could remember the incident of being lost in the mall—even though it hadn't actually happened. And it didn't stop there. Loftus explains: "They may start to remember a little bit about it. But when they come back a week later, they're starting to remember more. Maybe they'll talk about the older woman, who rescued them." Over time, more and more detail crept into the false memory: "The old lady was wearing this crazy hat"; "I had my favorite toy with me"; "My mom was so mad."

So not only was it possible to implant false new memories in the brain, but people embraced and embellished them, unknowingly weaving fantasy into the fabric of their identity.

We're all susceptible to this memory manipulation—even Loftus herself. As it turns out, when Elizabeth was a child, her mother had drowned in a swimming pool. Years later, a conversation with a relative brought out an extraordinary fact: that Elizabeth had been the one to find her mother's body in the pool. That news came as a shock to her; she hadn't known that, and in fact she didn't believe it. But, she describes, "I went home from that birthday and I started to think: maybe I did. I started to think about other things that I did remember—like when the firemen came, they gave me oxygen. Maybe I needed the oxygen 'cause I was so upset that I found the body?" Soon, she could visualize her mother in the swimming pool.

But then her relative called to say he had made a mistake. It wasn't the young Elizabeth after all who had found the body. It had been Elizabeth's aunt. And that's how Loftus had the opportunity to experience what it was like to possess her own false memory, richly detailed and deeply felt.

Our past is not a faithful record. Instead it's a reconstruction, and sometimes it can border on mythology. When we review our life memories, we should do so with the awareness that not all the details are accurate. Some came from stories that people told us about ourselves; others were filled in with what we

MEMORY OF THE FUTURE

Henry Molaison suffered his first major epileptic seizure on his fifteenth birthday. From there, the seizures grew more frequent. Faced with a future of violent convulsions, Henry underwent an experimental surgery—one which removed the middle part of the temporal lobe (including the hippocampus) on both sides of his brain. Henry was cured of the seizures, but with a dire side effect: for the rest of his life, he was unable to establish any new memories.

But the story doesn't end there. Beyond his inability to form new memories, he was also unable to imagine the future.

Picture what it would be like to go to the beach tomorrow. What do you conjure up? Surfers and sandcastles? Crashing waves? Rays of sun breaking through clouds? If you were to ask Henry what he might imagine, a typical response might be, "all I can come up with is the color blue." His misfortune reveals something about the brain mechanisms that underlie memory: their purpose is not simply to record what has gone before but to allow us to project forward into the future. To imagine tomorrow's experience at the beach, the hippocampus, in particular, plays a key role in assembling an imagined future by recombining information from our past.

thought must have happened. So if your answer to who you are is based simply on your memories, that makes your identity something of a strange, ongoing, mutable narrative.

THE AGING BRAIN

Today we're living longer than at any point in human history—and this presents challenges for maintaining brain health. Diseases like Alzheimer's and Parkinson's attack our brain tissue, and with it, the essence of who we are.

But here's the good news: in the same way that your environment and behavior shape your brain when you're younger, they are just as important in your later years.

Across the US, more than 1,100 nuns, priests, and brothers have been taking part in a unique research project—The Religious Orders Study—to explore the effects of aging on the brain. In particular the study is interested in teasing out the risk factors for Alzheimer's disease, and it includes subjects, aged sixty-five and over, who are symptom-free and don't exhibit any measurable signs of disease.

In addition to being a stable group that can be easily tracked down each year for regular tests, the religious orders share a similar lifestyle, including nutrition and living standards. This allows for fewer so-called "confounding factors," or differences, that might arise in the wider population, like diet or socioeconomic status or education—all of which could interfere with the study results.

Data collection began in 1994. So far, Dr. David Bennett and his team at Rush University in Chicago have collected over 350 brains. Each one is carefully preserved, and examined for microscopic evidence of age-related brain diseases. And that's only half the study: the other half involves the collection of in-depth data on each participant while they're alive. Every year, everyone in the study undergoes a battery of tests, ranging from psychological and cognitive appraisals to medical, physical, and genetic tests.

When the team began their research, they expected to find a clear-cut link between cognitive decline and the three diseases that are the most common causes of dementia: Alzheimer's, stroke and Parkinson's. Instead, here's what they found: having brain tissue that was being riddled with the ravages of Alzheimer's disease didn't necessarily mean a person would experience cognitive problems. Some people were dying with a full-blown Alzheimer's pathology without having cognitive loss. What was going on?

The team went back to their substantial data sets for clues. Bennett found that psychological and experiential factors determined whether there was loss of cognition. Specifically, cognitive exercise—that is, activity that keeps the brain active, like crosswords, reading, driving, learning new skills, and having responsibilities—was protective. So were social activity, social networks and interactions, and physical activity.

On the flip side, they found that negative psychological factors like loneliness, anxiety, depression, and proneness to psychological distress were related to more rapid cognitive decline. Positive traits like conscientiousness, purpose in life, and keeping busy were protective.

The participants with diseased neural tissue—but no cognitive symptoms—have built up what is known as "cognitive reserve." As areas of brain tissue have degenerated, other areas have been well exercised, and therefore have compensated or taken over those functions. The more we keep our brains cognitively fit—typically by challenging them with difficult and novel tasks, including social interaction—the more the neural networks build new roadways to get from A to B.

Think of the brain like a toolbox. If it's a good toolbox, it will contain all the tools you need to get a job done. If you need to disengage a bolt, you might fish out a ratchet; if you don't have access to the ratchet, you'll pull out a wrench; if the wrench is missing you might try a pair of pliers. It's the same concept in a cognitively fit brain: even if many pathways degenerate because of disease, the brain can retrieve other solutions.

The nuns' brains demonstrate that it's possible to protect our brains, and to help hold on to who we are for as long as possible. We can't stop the process of aging, but by practicing all the skills in our cognitive toolbox, we may be able to slow it down.

I AM SENTIENT

When I think about who I am, there's one aspect above all that can't be ignored: I am a sentient being. I experience my existence. I feel like I'm here, looking out on the world through these eyes, perceiving this Technicolor show from my own center stage. Let's call this feeling consciousness or awareness.

Scientists often debate the detailed definition of consciousness, but it's easy enough to pin down what we're talking about with the help of a simple comparison: when you're awake you have consciousness, and when you're in deep sleep you don't. That distinction gives us an inroad for a simple question: what is the difference in brain activity between those two states?

One way to measure that is with electroencephalography (EEG), which captures a summary of billions of neurons firing by picking up weak electrical signals on the outside of the skull. It's a bit of a crude technique; sometimes it's compared to trying to understand the rules of baseball by holding a microphone against the outside of a baseball stadium. Nonetheless, EEG can offer immediate insights into the differences between the waking and sleeping states.

When you're awake, your brain waves reveal that your billions of neurons are engaged in complex exchanges with one another: think of it like thousands of individual conversations in the ball-game crowd.

When you go to sleep, your body seems to shut down. So it's a natural assumption that the neuronal

THE MIND–BODY PROBLEM

Conscious awareness is one of the most baffling puzzles of modern neuroscience. What is the relationship between our mental experience and our physical brains?

The philosopher René Descartes assumed that an immaterial soul exists separately from the brain. His speculation was that sensory input feeds into the pineal gland, which serves as the gateway to the immaterial spirit. (He most likely chose the pineal gland simply because it sits on the brain's midline, while most other brain features are doubled, one on each hemisphere.)

The idea of an immaterial soul is easy to imagine; however, it's difficult to reconcile with neuroscientific evidence. Descartes never got to wander a neurology ward. If he had, he would have seen that when brains change, people's personalities change. Some kinds of brain damage make people depressed. Other changes make them manic. Others adjust a person's religiosity, sense of humor, or appetite for gambling. Others make a person indecisive, delusional, or aggressive. Hence the difficulty in the framework that the mental is separable from the physical.

As we'll see, modern neuroscience works to tease out the relationship of detailed neural activity to specific states of consciousness. It's likely that a full understanding of consciousness will require new discoveries and theories; our field is still quite young.

stadium quiets down. But in 1953 it was discovered that such an assumption is incorrect: the brain is just as active at night as during the day. During sleep, neurons simply coordinate with one another differently, entering a more synchronized, rhythmic state. Imagine the crowd at the stadium doing an incessant Mexican wave, around and around.

As you can imagine, the complexity of the discussion in a stadium is much richer when thousands of discrete conversations are playing out. In contrast, when the crowd is entrained in a bellowing wave, it's a less intellectual time.

So who you are at any given moment depends on the detailed rhythms of your neuronal firing. During the day, the conscious you emerges from that integrated neural complexity. At night, when the interaction of your neurons changes just a bit, you disappear. Your loved ones have to wait until the next morning, when your neurons let the wave die and work themselves back into their complex rhythm. Only then do you return.

So who you are depends on what your neurons are up to, moment by moment.

BRAINS ARE LIKE SNOWFLAKES

After I finished graduate school, I had the opportunity to work with one of my scientific heroes, Francis Crick. By the time I met him, he had turned his efforts to addressing the problem of consciousness. The chalkboard in his office contained a great deal of writing; what always struck me was that one word was written in the middle much larger than the rest. That word was "meaning." We know a lot about the mechanics of neurons and networks and brain regions—but we don't know why all those signals coursing around in there mean anything to us. How can the matter of our brains cause us to care about anything?

The meaning problem is not yet solved. But here's what I think we can say: the meaning of something to you is all about your webs of associations, based on the whole history of your life experiences.

Imagine I were to take a piece of cloth, put some colored pigments on it, and display it to your visual system. Is that likely to trigger memories and fire up your imagination? Well, probably not, because it's just a piece of cloth, right?

But now imagine that those pigments on a cloth are arranged into a pattern of a national flag. Almost certainly that sight will trigger something for you—but the specific meaning is unique to your history of experiences. You don't perceive objects as they are. You perceive them as you are.

Each of us is on our own trajectory—steered by our genes and our experiences—and as a result every brain has a different internal life. Brains are as unique as snowflakes.

As your trillions of new connections continually form and re-form, the distinctive pattern means that no one like you has ever existed, or will ever exist again. The experience of your conscious awareness, right now, is unique to you.

And because the physical stuff is constantly changing, we are too. We're not fixed. From cradle to grave, we are works in progress.

Your interpretation of physical objects has everything to do with the historical trajectory of your brain—and little to do with the objects themselves. These two rectangles contain nothing but arrangements of color. A dog would appreciate no meaningful difference between them. Whatever reaction you have to these is all about you, not them.

2

WHAT IS REALITY?

How does the biological wetware of the brain give rise to our experience: the sight of emerald green, the taste of cinnamon, the smell of wet soil? What if I told you that the world around you, with its rich colors, textures, sounds, and scents is an illusion, a show put on for you by your brain? If you could perceive reality as it really is, you would be shocked by its colorless, odorless, tasteless silence. Outside your brain, there is just energy and matter. Over millions of years of evolution the human brain has become adept at turning this energy and matter into a rich sensory experience of being in the world. How?

THE ILLUSION OF REALITY

From the moment you awaken in the morning, you're surrounded with a rush of light and sounds and smells. Your senses are flooded. All you have to do is show up every day, and without thought or effort, you are immersed in the irrefutable reality of the world.

But how much of this reality is a construction of your brain, taking place only inside your head?

Consider the rotating snakes, below. Although nothing is actually moving on the page, the snakes appear to be slithering. How can your brain perceive motion when you know that the figure is fixed in place?

Nothing moves on the page, but you perceive motion. Rotating Snakes illusion by Akiyoshi Kitaoka.

Compare the color of the squares marked A and B. Checkerboard illusion by Edward Adelson.

Or consider the checkerboard above.

Although it doesn't look like it, the square marked A is exactly the same color as the square marked B. Prove this to yourself by covering up the rest of the picture. How can the squares look so different, even though they're physically identical?

Illusions like these give us the first hints that our picture of the external world isn't necessarily an accurate representation. Our perception of reality has less to do with what's happening out there, and more to do with what's happening inside our brain.

YOUR EXPERIENCE OF REALITY

It feels as though you have direct access to the world through your senses. You can reach out and touch the material of the physical world—like this book or the chair you're sitting on. But this sense of touch is not a direct experience. Although it feels like the touch is happening in your fingers, in fact it's all happening in the mission control center of the brain. It's the same across all your sensory experiences. Seeing isn't happening in your eyes; hearing isn't taking place in your ears; smell isn't happening in your nose. All of your sensory experiences are taking place in storms of activity within the computational material of your brain.

Here's the key: the brain has no access to the world outside. Sealed within the dark, silent chamber of your skull, your brain has never directly experienced the external world, and it never will.

Instead, there's only one way that information from out there gets into the brain. Your sensory organs—your eyes, ears, nose, mouth, and skin—act as interpreters. They detect a motley crew of information sources (including photons, air compression waves, molecular concentrations, pressure, texture, temperature) and translate them into the common currency of the brain: electrochemical signals.

These electrochemical signals dash through dense networks of neurons, the main signaling cells of the brain. There are a hundred billion neurons in the human brain, and each neuron sends tens or hundreds of electrical pulses to thousands of other neurons every second of your life.

Everything you experience—every sight, sound,

smell—rather than being a direct experience, is an electrochemical rendition in a dark theater.

How does the brain turn its immense electrochemical patterns into a useful understanding of the world? It does so by comparing the signals it receives from the different sensory inputs, detecting patterns that allow it to make its best guesses about what's "out there." Its operation is so powerful that its work seems effortless. But let's take a closer look.

Let's begin with our most dominant sense: vision. The act of seeing feels so natural that it's hard to appreciate the immense machinery that makes it happen. About a third of the human brain is dedicated to the mission of vision, to turning raw photons of light into our mother's face, or our loving pet, or the couch we're about to nap on. To unmask what's happening under the hood, let's turn to the case of a man who lost his vision, and then was given the chance to get it back.

I WAS BLIND BUT NOW I SEE

Mike May lost his sight at the age of three and a half. A chemical explosion scarred his corneas, leaving his eyes with no access to photons. As a blind man, he became successful in business, and also became a championship Paralympic skier, navigating the slopes by sound markers.

Then, after over forty years of blindness, Mike learned about a pioneering stem-cell treatment that

SENSORY TRANSDUCTION

Biology has discovered many ways to convert information from the world into electrochemical signals. Just a few of the translation machines that you own: hair cells in the inner ear, several types of touch receptors in the skin, taste buds in the tongue, molecular receptors in the olfactory bulb, and photoreceptors at the back of the eye.

Signals from the environment are translated into electrochemical signals carried by brain cells. It is the first step by which the brain taps into information from the world outside the body. The eyes convert (or transduce) photons into electrical signals. The mechanisms of the inner ear convert vibrations in the density of the air into electrical signals. Receptors on the skin (and also inside the body) convert pressure, stretch, temperature, and noxious chemicals into electrical signals. The nose converts drifting odor molecules, and the tongue converts taste molecules to electrical signals. In a city with visitors from all over the world, foreign money must be translated into a common currency before meaningful transactions can take place. And so it is with the brain. It's fundamentally cosmopolitan, welcoming travelers from many different origins.

One of neuroscience's unsolved puzzles is known as the *"binding problem"*: how is the brain able to produce a single, unified picture of the world, given that vision is processed in one region, hearing in another, touch in another, and so on? While the problem is still unsolved, the common currency among neurons—as well as their massive interconnectivity—promises to be at the heart of the solution.

could repair the physical damage to his eyes. He decided to undertake the surgery; after all, the blindness was only the result of his unclear corneas, and the solution was straightforward.

But something unexpected happened. Television cameras were on hand to document the moment the bandages came off. Mike describes the experience when the physician peeled back the gauze: "There's this whoosh of light and bombarding of images on to my eye. All of a sudden you turn on this flood of visual information. It's overwhelming."

Mike's new corneas were receiving and focusing light just as they were supposed to. But his brain could not make sense of the information it was receiving. With the news cameras rolling, Mike looked at his children and smiled at them. But inside he was petrified, because he couldn't tell what they looked like, or which was which. "I had no face recognition whatsoever," he recalls.

In surgical terms, the transplant had been a total success. But from Mike's point of view, what he was experiencing couldn't be called vision. As he summarized it: "my brain was going 'oh my gosh.'"

With the help of his doctors and family, he walked out of the exam room and down the hallway, casting his gaze toward the carpet, the pictures on the wall, the doorways. None of it made sense to him. When he was placed in the car to go home, Mike set his eyes on the cars, buildings, and people whizzing by, trying unsuccessfully to understand what he was seeing. On the freeway, he recoiled when it looked like they were going to smash into a large rectangle in front of them. It turned out to be a highway sign, which they passed under. He had no sense of what objects were, nor of their depth. In fact, postsurgery, Mike found skiing more difficult than he had as a blind man. Because of his depth perception difficulties, he had a hard time telling the difference between people, trees, shadows, and holes. They all appeared to him simply like dark things against the white snow.

The lesson that surfaces from Mike's experience is that the visual system is not like a camera. It's not as though seeing is simply about removing the lens cap. For vision, you need more than functioning eyes.

In Mike's case, forty years of blindness meant that the territory of his visual system (what we would normally call the visual cortex) had been largely taken over by his remaining senses, such as hearing and touch. That impacted his brain's ability to weave together all the signals it needed to have sight. As we will see, vision emerges from the coordination of billions of neurons working together in a particular, complex symphony.

Today, fifteen years after his surgery, Mike still has a difficult time reading words on paper and the expressions on people's faces. When he needs to make better sense of his imperfect visual perception, he uses his other senses to cross-check the information: he touches, he lifts, he listens. This comparison across the senses is something we all did at a much younger age, when our brains were first making sense of the world.

SEEING REQUIRES MORE THAN THE EYES

When babies reach out to touch what's in front of them, it's not only to learn about texture and shape. These actions are also necessary for learning how to see. While it sounds strange to imagine that the movement of our bodies is required for vision, this concept was elegantly demonstrated with two kittens in 1963.

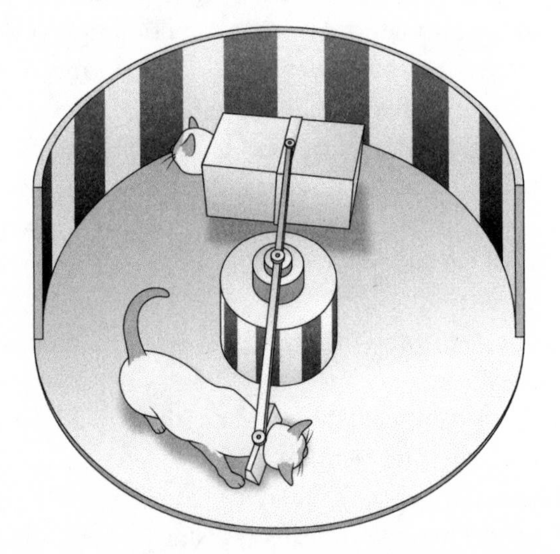

Inside a cylinder with vertical stripes, one kitten walked while the other was carried. Both received exactly the same visual input, but only the one who walked itself—the one able to match its own movements to changes in visual input—learned to see properly.

Richard Held and Alan Hein, two researchers at MIT, placed two kittens into a cylinder ringed in vertical stripes. Both kittens got visual input from moving around inside the cylinder. But there was a critical difference in their experiences: the first kitten was walking of its own accord, while the second kitten was riding in a gondola attached to a central axis. Because of this setup, both kittens saw exactly the same thing: the stripes moved at the same time and at the same speed for both. If vision were just about the photons hitting the eyes, their visual systems should develop identically. But here was the surprising result: only the kitten that was using its body to do the moving developed normal vision. The kitten riding in the gondola never learned to see properly; its visual system never reached normal development.

Vision isn't about photons that can be readily interpreted by the visual cortex. Instead it's a whole body experience. The signals coming into the brain can only be made sense of by training, which requires cross-referencing the signals with information from our actions and sensory consequences. It's the only way our brains can come to interpret what the visual data actually means.

If from birth you were unable to interact with the world in any way, unable to work out through feedback what the sensory information meant, in theory you would never be able to see. When babies hit the bars of their cribs and chew their toes and play with their blocks, they're not simply exploring—

they're training up their visual systems. Entombed in darkness, their brains are learning how the actions sent out into the world (turn the head, push this, let go of that) change the sensory input that returns. As a result of extensive experimentation, vision becomes trained up.

VISION FEELS EFFORTLESS BUT IT'S NOT

Seeing feels so effortless that it's hard to appreciate the effort the brain exerts to construct it. To lift the lid a little on the process, I flew to Irvine, California, to see what happens when my visual system doesn't receive the signals it expects.

Dr. Alyssa Brewer at the University of California is interested in understanding how adaptable the brain is. To that end, she outfits participants with prism goggles that flip the left and right sides of the world—and she studies how the visual system copes with it.

On a beautiful spring day, I strapped on the prism goggles. The world flipped—objects on the right now appeared on my left, and vice versa. When trying to figure out where Alyssa was standing, my visual system told me one thing, while my hearing told me another. My senses weren't matching up. When I reached out to grab an object, the sight of my own hand didn't match the position claimed by my muscles. Two minutes into wearing the goggles, I was sweating and nauseated.

Although my eyes were functioning and taking in the world, the visual data stream wasn't consistent with my other data streams. This spelled hard work for my brain. It was like I was learning to see again for the first time.

I knew that wearing the goggles wouldn't stay that difficult forever. Another participant, Brian Barton, was also wearing prism goggles—and he had been wearing them for a full week. Brian didn't seem to be on the brink of vomiting, as I was. To compare our levels of adaptation, I challenged him to a baking competition. The contest would require us to break eggs into a bowl, stir in cupcake mix, pour the batter into cupcake trays, and put the trays in the oven.

It was no contest: Brian's cupcakes came out of the oven looking normal, while most of my batter ended up dried onto the counter or baked in smears across the baking tray. Brian could navigate his world without much trouble, while I had been rendered inept. I had to struggle consciously through every move.

Wearing the goggles allowed me to experience the normally hidden effort behind visual processing. Earlier that morning, just before putting on the goggles, my brain could exploit its years of experience with the world. But after a simple reversal of one sensory input, it couldn't any longer.

To progress to Brian's level of proficiency, I knew I would need to continue interacting with the world for many days: reaching out to grab objects, following the direction of sounds, attending to the positions of my limbs. With enough practice, my brain would get trained up by a continual cross-referencing between the senses, just the way that Brian's brain had been doing for seven days. With training, my neural networks would figure out how various data streams entering into the brain matched up with other data streams.

Brewer reports that after a few days of wearing the goggles, people develop an internal sense of a new left and an old left, and a new right and an old right. After a week, they can move around normally, the way Brian could, and they lose the concept of which right and left were the old ones and new ones. Their spatial map of the world alters. By two weeks into the task, they can write and read well, and they walk and reach with the proficiency of someone without goggles. In that short time span, they master the flipped input.

The brain doesn't really care about the details of the input; it simply cares about figuring out how to most efficiently move around in the world and get what it needs. All the hard work of dealing with the low-level signals is taken care of for you. If you ever get a chance to wear prism goggles, you should. It exposes how much effort the brain goes through to make vision seem effortless.

SYNCHRONIZING THE SENSES

So we've seen that our perception requires the brain to compare different streams of sensory data against one another. But there's something which makes this sort of comparison a real challenge, and that is the issue of timing. All of the streams of sensory data—vision, hearing, touch, and so on—are processed by the brain at different speeds.

Consider sprinters at a racetrack. It appears that they get off the blocks at the instant the gun fires. But it's not actually instantaneous: if you watch them in slow motion, you'll see the sizable gap between the bang and the start of their movement—almost two tenths of a second. (In fact, if they move off the blocks before that duration, they're disqualified—they've "jumped the gun.") Athletes train to make this gap as small as possible, but their biology imposes fundamental limits: the brain has to register the sound, send signals to the motor cortex, and then down the spinal cord to the muscles of the body. In a sport where thousandths of a second can be the difference between winning and losing, that response seems surprisingly slow.

Could the delay be shortened if we used, say, a flash instead of a pistol to start the racers? After all, light travels faster than sound—so wouldn't that allow them to break off the blocks faster?

I gathered up some fellow sprinters to put this to the test. In the top photograph on page 52, we are triggered by a flash of light; in the bottom photo we're triggered by the gun.

We responded more slowly to the light. At first this may seem counterintuitive, given the speed of light in the outside world. But to understand what's happening we need to look at the speed of information processing on the inside. Visual data goes through more complex processing than auditory data. It takes longer for signals carrying flash information to work their way through the visual system than for bang signals to work through the auditory system. We were able to respond to the light at 190 milliseconds, but to a bang at only 160 milliseconds. That's why a pistol is used to start sprinters.

But here's where it gets strange. We've just seen that the brain processes sounds more quickly than sights. And yet take a careful look at what happens when you clap your hands in front of you. Try it. Everything seems synchronized. How can that be, given that sound

Sprinters can break off the blocks more quickly to a bang (bottom panel) than to a flash (top panel).

is processed more quickly? What it means is that your perception of reality is the end result of fancy editing tricks: the brain hides the difference in arrival times. How? What it serves up as reality is actually a delayed version. Your brain collects up all the information from the senses before it decides upon a story of what happens.

These timing difficulties aren't restricted to hearing and seeing: each type of sensory information takes a different amount of time to process. To complicate things even more, even within a sense there are time differences. For example, it takes longer for signals to reach your brain from your big toe than it does from your nose. But none of this is obvious to your perception: you collect up all the signals first, so that everything seems synchronized. The strange consequence of all this is that you live in the past. By the time you think the moment occurs, it's already long gone. To synchronize the incoming information from the senses, the cost is that our conscious awareness lags behind the physical world. That's the unbridgeable gap between an event occurring and your conscious experience of it.

WHEN THE SENSES ARE CUT OFF, DOES THE SHOW STOP?

Our experience of reality is the brain's ultimate construction. Although it's based on all the streams of data from our senses, it's not dependent on them. How

THE BRAIN IS LIKE A CITY

Just like a city, the brain's overall operation emerges from the networked interaction of its innumerable parts. There is often a temptation to assign a function to each region of the brain, in the form of "this part does that." But despite a long history of attempts, brain function cannot be understood as the sum of activity in a collection of well-defined modules.

Instead, think of the brain as a city. If you were to look out over a city and ask "where is the economy located?" you'd see there's no good answer to the question. Instead, the economy emerges from the interaction of all the elements—from the stores and the banks to the merchants and the customers.

And so it is with the brain's operation: it doesn't happen in one spot. Just as in a city, no neighborhood of the brain operates in isolation. In brains and in cities, everything emerges from the interaction between residents, at all scales, locally and distantly. Just as trains bring materials and textiles into a city, which become processed into the economy, so the raw electrochemical signals from sensory organs are transported along superhighways of neurons. There the signals undergo processing and transformation into our conscious reality.

do we know? Because when you take it all away, your reality doesn't stop. It just gets stranger.

On a sunny San Francisco day, I took a boat across the chilly waters to Alcatraz, the famous island prison. I was going to see a particular cell called the Hole. If you broke the rules in the outside world, you were sent to Alcatraz. If you broke the rules in Alcatraz, you were sent to the Hole.

I entered the Hole and closed the door behind me. It's about ten by ten feet. It was pitch-black: not a photon of light leaks in from anywhere. Sounds are cut off completely. In here, you are left utterly alone with yourself.

What would it be like to be locked in here for hours, or for days? To find out, I spoke to a surviving inmate who had been here. Armed robber Robert Luke—known as Cold Blue Luke—was sent to the Hole for twenty-nine days for smashing up his cell. Luke described his experience: "The dark Hole was a bad place. Some guys couldn't take that. I mean, they were in there and in a couple of days they were banging their head on the wall. You didn't know how you would act when you got in there. You didn't want to find out."

Completely isolated from the outside world, with no sound and no light, Luke's eyes and ears were completely starved of input. But his mind didn't abandon the notion of an outside world. It just continued to make one up. Luke describes the experience: "I remember going on these trips. One I used to remember was flying a kite. It got pretty real. But they were all in my head." Luke's brain continued to see.

Such experiences are common among prisoners in solitary confinement. Another resident of the Hole described seeing a spot of light in his mind's eye; he would expand that spot into a television screen and watch TV. Deprived of new sensory information, prisoners said they went beyond daydreaming: instead, they spoke of experiences that seemed completely real. They didn't just imagine pictures, they saw.

This testimony illuminates the relationship between the outside world and what we take to be reality. How can we understand what was going on with Luke? In the traditional model of vision, perception results from a procession of data that begins from the eyes and ends with some mysterious end point in the brain. But despite the simplicity of that assembly-line model of vision, it's incorrect.

In fact, the brain generates its own reality, even before it receives information coming in from the eyes and the other senses. This is known as the internal model.

The basis of the internal model can be seen in the brain's anatomy. The thalamus sits between the eyes at the front of the head and the visual cortex at the back of the head. Most sensory information connects through here on its way to the appropriate region of the cortex. Visual information goes to the visual cortex, so there are a huge number of connections going from the thalamus into the visual cortex. But here's the surprise: there are ten times as many going in the opposite direction.

Detailed expectations about the world—in other words, what the brain "guesses" will be out there—are being transmitted by the visual cortex to the thalamus. The thalamus then compares what's coming in from the eyes. If that matches the expectations ("when I turn my head I should see a chair there"), then very little activity goes back to the visual system. The thalamus simply reports on differences between what the eyes are reporting, and what the brain's internal model has predicted. In other words, what gets sent back to the visual cortex is what fell short in the expectation (also known as the "error"): the part that wasn't predicted away.

So at any moment, what we experience as seeing relies less on the light streaming into our eyes, and more on what's already inside our heads.

And that's why Cold Blue Luke sat in a pitch-black cell having rich visual experiences. Locked in the Hole, his senses were providing his brain with no new input, so his internal model was able to run free, and he experienced vivid sights and sounds. Even when brains are unanchored from external data, they continue to generate their own imagery. Remove the world and the show still goes on.

You don't have to be locked up in the Hole to experience the internal model. Many people find great pleasure in sensory deprivation chambers—dark pods in which they float in salty water. By removing the anchor of the external world, they let the internal world fly free.

And of course you don't have to go far to find your own sensory deprivation chamber. Every night when you go to sleep you have full, rich, visual experiences. Your eyes are closed, but you enjoy the lavish and colorful world of your dreams, believing the reality of every bit of it.

SEEING OUR EXPECTATIONS

When you walk down a city street, you seem to automatically know what things are without having to work out the details. Your brain makes assumptions about what you're seeing based on your internal model, built up from years of experience of walking other city streets. Every experience you've had contributes to the internal model in your brain.

Instead of using your senses to constantly rebuild your reality from scratch every moment, you're comparing sensory information with a model that the brain has already constructed: updating it, refining it, correcting it. Your brain is so expert at this task that you're normally unaware of it. But sometimes, under certain conditions, you can see the process at work.

Try taking a plastic mask of a face, the type you wear on Halloween. Now rotate around so you're looking at the hollow backside. You know it's hollow. But despite this knowledge, you often can't help but see the face as though it's coming out at you. What you experience is not the raw data hitting your eyes, but instead your internal model—a model which has been trained on a lifetime of faces that stick out. The hollow mask illusion reveals the strength of your expectations in what you see. (Here's an easy way to demonstrate the hollow mask illusion to yourself: stick your face into fresh snow and take a photo of the impression. The resulting picture looks to your brain like a 3-D snow sculpture that's sticking out.)

When you're confronted with the hollow side of a mask (right), it still looks like it's coming toward you. What we see is strongly influenced by our expectations.

It's also your internal model that allows the world out there to remain stable—even when you're moving. Imagine you were to see a cityscape that you really wanted to remember. So you take out your cell phone to capture a video. But instead of smoothly panning your camera across the scene, you decide to move it around exactly as your eyes move around. Although you're not generally aware of it, your eyes jump around about four times a second, in jerky movements called saccades. If you were to film this way, it wouldn't take you long to discover that this is no way to take a video: when you play it back, you'd find that your rapidly lurching video is nauseating to watch.

So why does the world appear stable to you when you're looking at it? Why doesn't it appear as jerky and nauseating as the poorly filmed video? Here's why: your internal model operates under the assumption that the world outside is stable. Your eyes are not like video cameras—they simply venture out to find more details to feed into the internal model. They're not like camera lenses that you're seeing through; they're gathering bits of data to feed the world inside your skull.

OUR INTERNAL MODEL IS LOW RESOLUTION BUT UPGRADABLE

Our internal model of the outside world allows us to get a quick sense of our environment. And that is its primary function—to navigate the world. What's not always obvious is how much of the finer detail the brain leaves out. We have the illusion that we're taking in the world around us in great detail. But as an experiment from the 1960s shows, we aren't.

Russian psychologist Paul Yarbus devised a way to track people's eyes as they took in a scene for the first time. Using the painting *The Unexpected Visitor* by Ilya Repin, he asked subjects to take in its details over three minutes, and then to describe what they had seen after the painting was hidden away.

In a rerun of his experiment, I gave participants time to take in the painting, time for their brains to build an internal model of the scene. But how detailed was that model? When I asked the participants questions, everyone who had seen the painting thought they knew what was in it. But when I asked about specifics, it became clear that their brains hadn't filled in most of the details. How many paintings were on the walls? What was the furniture in the room? How many children? Carpet or wood on the floor? What was the expression on the face of the unexpected visitor? The lack of answers revealed that people had taken in only a very cursory sense of the scene. They were surprised to discover that even with a low-resolution internal model, they still had the impression that everything had been seen. Later, after the questions, I gave them a chance to look again at the painting to seek out some of the answers. Their eyes sought out the information and incorporated it for a new, updated internal model.

This isn't a failure of the brain. It doesn't try to produce a perfect simulation of the world. Instead, the internal model is a hastily drawn approximation—as long as the brain knows where to go to look for the finer points, more details are added on a need-to-know basis.

So why doesn't the brain give us the full picture? Because brains are expensive, energy-wise. Twenty percent of the calories we consume are used to power the brain. So brains try to operate in the most energy-efficient way possible, and that means processing only the minimum amount of information from our senses that we need to navigate the world.

We tracked eye movements as volunteers looked at *The Unexpected Visitor*, a painting by Ilya Repin. The white streaks show where their eyes went. Despite the coverage with eye movements, they retained almost none of the detail.

Neuroscientists weren't the first to discover that fixing your gaze on something is no guarantee of seeing it. Magicians figured this out long ago. By directing your attention, magicians perform sleight of hand in full view. Their actions should give away the game, but they can rest assured that your brain processes only small bits of the visual scene.

This all helps to explain the prevalence of traffic

accidents in which drivers hit pedestrians in plain view, or collide with cars directly in front of them. In many of these cases, the eyes are pointed in the right direction, but the brain isn't seeing what's really out there.

TRAPPED ON A THIN SLICE OF REALITY

We think of color as a fundamental quality of the world around us. But in the outside world, color doesn't actually exist.

When electromagnetic radiation hits an object, some of it bounces off and is captured by our eyes. We can distinguish between millions of combinations of wavelengths—but it is only inside our heads that any of this becomes color. Color is an interpretation of wavelengths, one that only exists internally.

And it gets stranger, because the wavelengths we're talking about involve only what we call "visible light," a spectrum of wavelengths that runs from red to violet. But visible light constitutes only a tiny fraction of the electromagnetic spectrum—less than one ten-trillionth of it. All the rest of the spectrum—including radio waves, microwaves, X-rays, gamma rays, cell phone conversations, Wi-Fi, and so on—all of this is flowing through us right now, and we're completely unaware of it. This is because we don't have any specialized biological receptors to pick up on these signals from other parts of the spectrum. The slice of reality that we can see is limited by our biology.

Each creature picks up on its own slice of reality. In the blind and deaf world of the tick, the signals it detects from its environment are temperature and body odor. For bats, it's the echolocation of air compression waves. For the black ghost knifefish, its experience of the world is defined by perturbations in electrical fields. These are the slices of their ecosystem that they can detect. No one is having an experience of the objective reality that really exists; each creature perceives only what it has evolved to perceive. But presumably, every creature assumes its slice of reality to be the entire objective world. Why would we ever stop to imagine there's something beyond what we can perceive?

So what does the world outside your head really "look" like? Not only is there no color, there's also no sound: the compression and expansion of air is picked up by the ears, and turned into electrical signals. The brain then presents these signals to us as mellifluous tones and swishes and clatters and jangles. Reality is also odorless: there's no such thing as smell outside our brains. Molecules floating through the air bind to receptors in our nose and are interpreted as different smells by our brain. The real world is not full of rich sensory events; instead, our brains light up the world with their own sensuality.

YOUR REALITY, MY REALITY

How do I know if my reality is the same as yours? For most of us it's impossible to tell, but there's a small fraction of the population whose perception of reality is measurably different from ours.

Consider Hannah Bosley. When she looks at letters of the alphabet, she has an internal experience of color. For her, it's self-evidently true that J is purple, or that T is red. Letters automatically and involuntarily trigger color experiences, and her associations never change. Her first name looks to her like a sunset, starting with yellow, fading into red, then to a color like clouds, and then back into red and to yellow. The name "Iain," in contrast, looks like vomit to her, although she's perfectly nice to people with that name.

Hannah is not being poetic or metaphorical—she has a perceptual experience known as synesthesia. Synesthesia is a condition in which senses (or in some cases concepts) are blended. There are many different kinds of synesthesia. Some taste words. Some see sounds as colors. Some hear visual motion. About 3 percent of the population has some form of synesthesia.

Hannah is just one of over 6,000 synesthetes I have studied in my lab; in fact, Hannah worked in my lab for two years. I study synesthesia because it's one of the few conditions in which it's clear that someone else's experience of reality is measurably different from mine. And it makes it obvious that how we perceive the world is not one-size-fits-all.

Synesthesia is the result of cross-talk between sensory areas of the brain, like neighboring districts with porous borders. Synesthesia shows us that even microscopic changes in brain wiring can lead to different realities.

Every time I meet someone who has this kind of experience, it's a reminder that from person to person—and from brain to brain—our internal experience of reality can be somewhat different.

BELIEVING WHAT OUR BRAINS TELL US

We all know what it is to have dreams at night, to have bizarre, unbidden thoughts that take us on journeys. Sometimes these are disturbing journeys we have to suffer through. The good news is that when we wake up, we are able to compartmentalize: that was a dream, and this is my waking life.

Imagine what it would be like if these states of your reality were more intertwined, and it were more difficult—or impossible—to distinguish one from the other. For about 1 percent of the population, that distinction can be difficult, and their realities can be overwhelming and terrifying.

Elyn Saks is a professor of law at the University of Southern California. She's smart and kind, and she's been sporadically experiencing schizophrenic episodes since she was sixteen years old. Schizophrenia is a disorder of her brain function, causing her to hear voices, or see things others don't see, or believe that other people are reading her thoughts. Fortunately, thanks to medication and weekly therapy sessions, Elyn has been able to lecture and teach at the law school for over twenty-five years.

I spoke with her at USC, and she gave me examples

of schizophrenic episodes she's had in the past. "I felt like the houses were communicating with me: You are special. You are especially bad. Repent. Stop. Go. I didn't hear these as words, but I heard them as thoughts put into my head. But I knew they were the houses' thoughts, and not my thoughts." In one incident, she believed that explosions were being set off in her brain, and she was afraid that this was going to hurt other people, not just her. At a different time in her life she held a belief that her brain was going to leak out of her ears and drown people.

Now, having escaped those delusions, she laughs and shrugs, wondering what it was all about.

It was about chemical imbalances in her brain that subtly changed the pattern of signals. A slightly different pattern, and one can suddenly be trapped inside a reality in which strange and impossible things unfold. When Elyn was inside a schizophrenic episode, it never struck her that something was strange. Why? Because she believed the narrative told by the sum of her brain chemistry.

I once read an old medical text in which schizophrenia was described as an intrusion of the dream state into the waking state. Although I don't often see it described that way anymore, it's an insightful way to understand what the experience would be like from the inside. The next time you see someone on a street corner talking to himself and acting out a narrative, remind yourself what it would be like if you couldn't distinguish your waking and sleeping states.

Elyn's experience is an inroad to understanding our own realities. When we're in the middle of a dream, it seems real. When we've misinterpreted a quick glance of something we've seen, it's hard to shake the feeling that we know the reality of what we saw. When we're recalling a memory that is, in fact, false, it's difficult to accept claims that it didn't really happen. Although it's impossible to quantify, accumulations of such false realities color our beliefs and actions in ways of which we can never be cognizant.

Whether she was in the thick of a delusion, or else aligned with the reality of the broader population, Elyn believed that what she was experiencing was really happening. For her, as with all of us, reality is a narrative played out inside the sealed auditorium of the cranium.

TIMEWARP

There's another facet of reality that we rarely stop to consider: our brain's experience of time can often be quite strange. In certain situations, our reality can seem to run more slowly or more quickly.

When I was eight years old I fell off the roof of a house, and the fall seemed to take a very long time. When I got to high school I learned physics and I calculated how long the fall actually took. It turns out it took eight-tenths of a second. So that set me off on a quest to understand something: why did it seem to take so long and what did this tell me about our perception of reality?

Up above the mountains, professional wingsuit flyer Jeb Corliss has experienced time distortion. It all began with a particular jump he'd done before. But on this day, he decided to aim for a target: a set of balloons to smash past with his body. Jeb recalls: "As I was coming in to hit one of those balloons, tied to a ledge of granite, I misjudged." He bounced off flat granite at what he estimates to be 120 miles per hour.

Because Jeb wingsuits professionally, the events this day were captured by a collection of cameras on the cliffs and on his body. In the video, one can hear the thump as Jeb hits the granite. He streaks past the cameras and keeps going, over the edge of the cliffside he's just scraped against.

And here's where Jeb's sense of time warped. As he describes it: "My brain split into two separate thought processes. One of the thought processes was just technical data. You've got two options: you cannot pull, so you go ahead and impact and basically die. Or, you can pull, get a parachute over your head and then bleed to death while you're waiting for rescue."

To Jeb these two separate thought processes felt like minutes of time: "It feels like you're operating so fast that your perception of everything else seems to slow down, and everything gets stretched. Time slows down and you get that feeling of slow motion."

He pulled his rip cord and careened to the ground having broken a leg, both ankles, and three toes. Six seconds elapsed between the instant Jeb hit the rock, and the moment he yanked the cord. But, just like my fall from the roof, that stretch seemed to him to have taken a longer time.

The subjective experience of time slowing has been reported in a variety of life-threatening experiences—for example, car accidents or muggings—as well as in events that involve seeing a loved one in danger, such as a child falling into a lake. All these reports are characterized by a sense that the events unfolded more slowly than normal, with rich details available.

When I fell off the roof, or when Jeb bounced off the cliff's lip, what happened inside our brains? Does time really slow down in frightening situations?

A few years ago, my students and I designed an experiment to address this open question. We induced extreme fear in people by dropping them from 150 feet in the air. In free fall. Backward.

In this experiment, participants fell with a digital display strapped to their wrists—a device we invented called the perceptual chronometer. They reported the numbers they were able to read on the device strapped to their wrists. If they really could see time in slow motion they would be able to read the numbers. But no one could.

So why do Jeb and I both recall our accidents as happening in slow motion? The answer appears to lie in the way our memories are stored.

In threatening situations, an area of the brain called the amygdala kicks into high gear, commandeering the resources of the rest of the brain and forcing everything to attend to the situation at hand. When the amygdala

MEASURING THE SPEED OF SIGHT: THE PERCEPTUAL CHRONOMETER

To test time perception in frightening situations, we dropped volunteers from 150 feet. I dropped myself three times; each time was equally terrifying. On the display, numbers are generated with LED lights. Every moment, the lights that are on go off, and those that are off turn on. At slow rates of alternation, participants have no trouble reporting the numbers. But at a slightly faster rate, the positive and negative images fuse together, making the numbers impossible to see. To determine whether participants could actually see in slower motion, we dropped people with the alternation rate just slightly higher than people could normally see. If they were actually seeing in slow motion—like Neo in *The Matrix*—they would have no trouble discriminating the numbers. If not, the rate at which they can perceive the numbers should be no different than when they were on the ground. The result? We dropped twenty-three volunteers, including myself. No one's in-flight performance was better than their ground-based performance. Despite initial hopes, we were not like Neo.

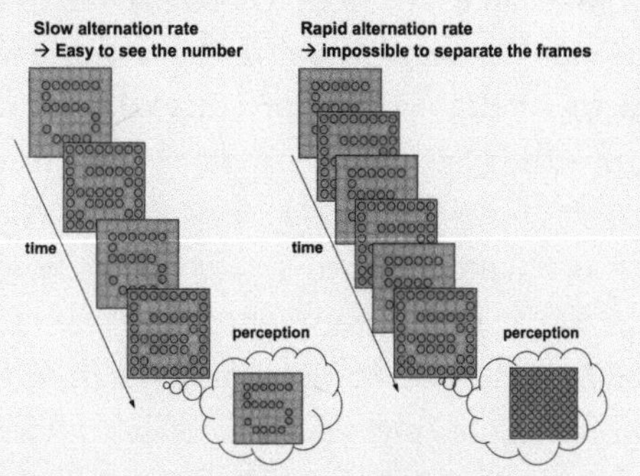

When the perceptual chronometer alternates numbers slowly, they can be read out. At a slightly higher alternation rate, they become impossible to read.

is in play, memories are laid down with far more detail and richness than under normal circumstances; a secondary memory system has been activated. After all, that's what memory is for: keeping track of important events, so that if you're ever in a similar situation, your brain has more information to try to survive. In other words, when things are life-threateningly scary, it's a good time to take notes.

The interesting side effect is this: your brain is not accustomed to that kind of density of memory (the hood was crumpling, the rearview mirror was falling off, the other driver looked like my neighbor Bob)—so when the events are replayed in your memory, your interpretation is that the event must have taken a longer time. In other words, it appears we don't actually experience terrifying accidents in slow motion; instead, the impression results from the way memories are read out. When we ask ourselves "What just happened?" the detail of memory tells us that it must have been in slow motion, even though it wasn't. Our time distortion is something that happens in retrospect, a trick of the memory that writes the story of our reality.

Now, if you've been in a life-threatening accident, you might insist that you were conscious of the slow-motion unfolding as it happened. But note: that's another trick about our conscious reality. As we saw above with the synchronizing of the senses, we're never actually present in the moment. Some philosophers suggest that conscious awareness is nothing but lots of fast memory querying: our brains are always asking

"What just happened? What just happened?" Thus, conscious experience is really just immediate memory.

As a side note, even after we published our research on this, some people still tell me that they know the event actually unfolded like a slow-motion movie. So I typically ask them whether the person next to them in the car was screaming like people do in slow-motion movies, with a low-pitched "noooooooo!" They have to allow that didn't happen. And that's part of why we think that perceptual time doesn't actually stretch out, a person's internal reality notwithstanding.

THE STORYTELLER

Your brain serves up a narrative—and each of us believes whatever narrative it tells. Whether you're falling for a visual illusion, or believing the dream you happen to be trapped in, or experiencing letters in color, or accepting a delusion as true during an episode of schizophrenia, we each accept our realities however our brains script them.

Despite the feeling that we're directly experiencing the world out there, our reality is ultimately built in the dark, in a foreign language of electrochemical signals. The activity churning across vast neural networks gets turned into your story of this, your private experience of the world: the feeling of this book in your hands, the light in the room, the smell of roses, the sound of others speaking.

Even more strangely, it's likely that every brain tells

a slightly different narrative. For every situation with multiple witnesses, different brains are having different private subjective experiences. With seven billion human brains wandering the planet (and trillions of animal brains), there's no single version of reality. Each brain carries its own truth.

So what is reality? It's like a television show that only you can see, and you can't turn it off. The good news is that it happens to be broadcasting the most interesting show you could ask for: edited, personalized, and presented just for you.

3

WHO IS IN CONTROL?

The cosmos turned out to be larger than we had ever imagined from gazing at the night sky. Similarly, the universe inside our heads extends far beyond the reach of our conscious experience. Today we are gaining the first glimpses of the enormity of this inner space. It seems to require very little effort for you to recognize a friend's face, drive a car, get a joke, or decide what to grab from the refrigerator—but in fact these things are possible only because of vast computations happening below your conscious awareness. At this moment, just like every moment of your life, networks in your brain are buzzing with activity: billions of electrical signals are racing along cells, triggering chemical pulses at trillions of connections between neurons. Simple acts are underpinned by a massive labor force of neurons. You remain blissfully unaware of all their activity, but your life is shaped and colored by what's happening under the hood: how you act, what matters to you, your reactions, your loves and desires, what you believe to be true and false. Your experience is the final output of these hidden networks. So who exactly is steering the ship?

CONSCIOUSNESS

It's morning. The streets of your neighborhood are quiet as the sun peeks above the horizon. In bedrooms all over your city, one by one, an astonishing event is taking place: human consciousness is flickering to life. The most complex object on our planet is becoming aware that it exists.

Just a while ago you, too, were in deep sleep. The biological material of your brain was the same then as it is now, but the activity patterns have slightly changed—so at this moment you're enjoying experiences. You're reading squiggles on a page and extracting meaning from them. You might be feeling sun on your skin and a breeze in your hair. You can think about the position of your tongue in your mouth or the feeling of your left shoe on your foot. Being awake, you're now aware of an identity, a life, needs, desires, plans. Now that the day has begun, you're ready to reflect on your relationships and goals, and guide your actions accordingly.

But how much control does your conscious awareness have over your daily operations?

Consider how you're reading these sentences. When

you pass your eyes over this page, you're mostly unaware of the rapid, ballistic jumps made by your eyes. Your eyes aren't moving smoothly across the page; instead, they dart from one fixed point to another. When your eyes are in the middle of a jump they're moving too fast to read. They only take in the text when you stop and fixate on one position, usually for twenty milliseconds or so at a time. We are not aware of these hops and jumps and stops and starts, because your brain goes to a lot of trouble to stabilize your perception of the outside world.

Reading gets even stranger when you consider this: as you read these words, their meaning flows from this sequence of symbols directly into your brain. To get a sense of the complexity of what's involved, try to read this same information in another language:

আপনার মস্তিষ্কের মধ্যে সরাসরি চিহ্ন এই ক্রম থেকে প্রবাহ অর্থ
эта азначае , патокі з сімвалаў непасрэдна ў ваш мозг
당신의 두뇌 에 직접 심볼 의 흐름을 의미

If you happen not to read Bengali, Belarussian, or Korean, then these letters appear to you simply as strange doodles. But once you've mastered reading a script (like this one), the act gives the illusion of being effortless: we are no longer aware that we are performing the arduous chore of deciphering squiggles. Your brain takes care of the work behind the scenes.

So who is in control? Are you the captain of your own boat, or do your decisions and actions have more to do with massive neural machinery operating out of sight? Does the quality of your everyday life have to do with your good decision making, or instead with dense jungles of neurons and the steady hum of innumerable chemical transmissions?

In this chapter, we'll discover that the conscious you is only the smallest part of the activity of your brain. Your actions, your beliefs and your biases are all driven by networks in your brain to which you have no conscious access.

THE UNCONSCIOUS BRAIN IN ACTION

Imagine we're sitting together in a coffee shop. As we're chatting, you notice me lift my cup of coffee to take a sip. The act is so unremarkable that it normally bears no mention unless I spill some on my shirt. But let's give credit where it's due: getting the cup to my mouth is no easy feat. The field of robotics still struggles to make this sort of task run without a hitch. Why? Because this simple act is underpinned by trillions of electrical impulses meticulously coordinated by my brain.

My visual system first scans the scene to pinpoint the cup in front of me, and my years of experience trigger memories of coffee in other situations. My frontal cortex deploys signals on a journey to my motor cortex, which precisely coordinates muscle contractions—throughout my torso, arm, forearm, and hand—so I can grasp the cup. As I touch the cup, my nerves carry

THE BRAIN FOREST

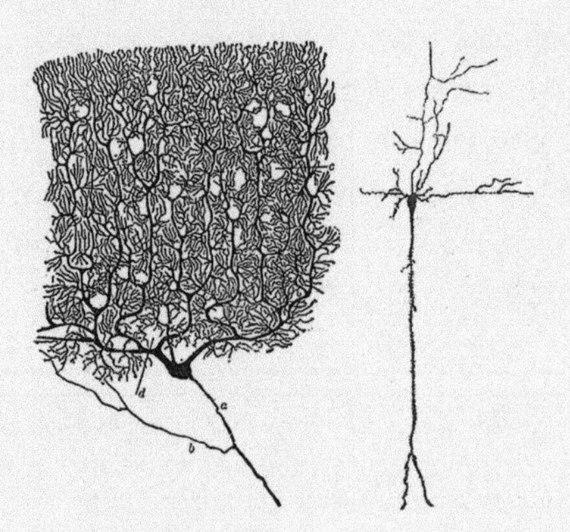

Beginning in 1887, the Spanish scientist Santiago Ramón y Cajal used his photography background to apply chemical stains to slices of brain tissue. This technique allowed individual cells in the brain, with all their branching beauty, to be seen. It began to become clear that the brain was a system of complexity for which we had no equivalent, and no language to capture it.

With the advent of mass-produced microscopes and new methods of staining cells, scientists began to describe—at least in general terms—the neurons that comprise our brains. These wondrous structures come in an intriguing variety of shapes and sizes, and are wired up in an impenetrably dense forest that scientists will be working to untangle for many decades to come.

back reams of information about the cup's weight, its position in space, its temperature, the slipperiness of the handle, and so on. As that information streams up the spinal cord and into the brain, compensating information streams back down, passing like fast-flowing traffic on a two-way road. This information emerges from a complex choreography between parts of my brain with names like basal ganglia, cerebellum, somatosensory cortex, and many more. In fractions of a second, adjustments are made to the force with which I'm lifting and the strength of my grip. Through intensive calculations and feedback, I adjust my muscles to keep the cup level as I smoothly move it on its long arc upward. I make micro-adjustments all along the way, and as it approaches my lips I tilt the cup just enough to extract some liquid without scalding myself.

It would take dozens of the world's fastest supercomputers to match the computational power required to pull off this feat. Yet I have no perception of this lightning storm in my brain. Although my neural networks are screaming with activity, my conscious awareness experiences something quite different. Something more like total obliviousness. The conscious me is engrossed in our conversation. So much so that I may even be shaping the airflow through my mouth while I'm lifting the cup, holding up my end of a complex conversation.

All I know is whether I get the coffee to my mouth or not. If executed perfectly, I'm likely to not even have noticed that I performed the act at all.

The unconscious machinery of our brains is at work all the time, but it runs so smoothly that we're typically unaware of its operations. As a result, it's often easiest to appreciate only when it stops working. What would it be like if we had to consciously think about simple actions that we normally take for granted, such as the seemingly straightforward act of walking? To find out, I went to speak with a man named Ian Waterman.

When Ian was nineteen years old, he suffered a rare type of nerve damage as a result of a fierce case of gastric flu. He lost the sensory nerves that tell the brain about touch, as well as the position of one's own limbs (known as proprioception). As a result, Ian could no longer manage any of the movements of his body automatically. Doctors told him that he would be confined to a wheelchair for the rest of his life, despite the fact that his muscles were fine. A person simply can't get around without knowledge of where his body is. Although we rarely pause to appreciate it, the feedback we get from the world and from our muscles makes possible the complex movements we manage every moment of the day.

Ian wasn't willing to let his condition confine him to a life without movement. So he gets up and goes, but the whole of his waking life requires him to think consciously about every movement his body makes. With no sense of awareness of where his limbs are, Ian has to move his body with focused, conscious determination. He uses his visual system to monitor the position of his limbs. As he walks, Ian leans his

PROPRIOCEPTION

Even with your eyes closed, you know where your limbs are: is your left arm up or down? Are your legs straight or bent? Is your back straight or slumped? This capacity to know the state of your muscles is called proprioception. Receptors in the muscles, tendons and joints provide information about the angles of your joints, as well as the tension and length of your muscles. Collectively, this gives the brain a rich picture of how the body is positioned and allows for fast adjustments.

You can experience your proprioception fail temporarily if you've ever attempted to walk after one of your legs has gone to sleep. Pressure on your squeezed sensory nerves has prevented the proper signals from being sent and received. Without a sense of the position of your own limbs, simple acts like cutting food, typing, or walking are almost impossible.

head forward to watch his limbs as best he can. To keep his balance, he compensates by making sure his arms are extended behind him. Because Ian can't feel his feet touch the floor, he must anticipate the exact distance of each step and land it with his leg braced. Every step he takes is calculated and coordinated by his conscious mind.

Having lost his ability to walk automatically, Ian is

highly cognizant of the miraculous coordination that most of us take for granted when going on a stroll. Everyone around him is moving around so fluidly and so seamlessly, he points out, that they're totally unaware of the amazing system that's managing that process for them.

If he is momentarily distracted, or an unrelated thought pops into his head, Ian is likely to fall. All distractions have to be tucked away while he concentrates on the smallest of details: the slope of the ground, the swing of his leg.

If you were to spend time with Ian for even a minute or two, it would immediately bring to light the exceeding complexity of the everyday acts we never even think to speak of: getting up, crossing the room, opening the door, reaching out to shake a hand. Despite first appearances, those acts aren't simple at all. So the next time you see a person walking, or jogging, or skateboarding, or riding a bicycle, take a moment to marvel not only at the beauty of the human body, but at the power of the unconscious brain that flawlessly orchestrates it. The intricate details of our most basic movements are animated by trillions of calculations, all buzzing along at a spatial scale smaller than you can see, and a complexity scale beyond what you can comprehend. We have yet to build robots that scratch the edges of human performance. And while a supercomputer racks up enormous energy bills, our brains work out what to do with baffling efficiency, using about the energy of a 60-watt lightbulb.

BURNING SKILLS INTO THE WIRING OF THE BRAIN

Neuroscientists often unlock clues into brain function by examining people who are specialized in some area. To that end, I went to meet Austin Naber, a ten-year-old boy with an extraordinary talent: he holds the children's world record for a sport known as cup stacking.

In quick, fluid movements impossible to follow with your eyes, Austin transforms a stacked column of plastic cups into a symmetrical display of three separate pyramids. Then, with both hands dashing, he telescopes the pyramids back down into two short columns, and then transmutes the columns into a single, tall pyramid, which is then collapsed into the original column of cups.

He does this all in five seconds. I tried it, and it took me forty-three seconds on my best run.

Watching Austin in action, you might expect his brain to be working overtime, burning an enormous amount of energy to coordinate these complex actions so quickly. To put this expectation to the test, I set out to measure his brain activity—and my own—during a head-to-head cup-stacking challenge. With the aid of researcher Dr. José Luis Contreras-Vidal, Austin and I were fitted with electrode caps to measure the electrical activity caused by populations of neurons beneath the skull. The brain waves measured by the electroencephalogram (EEG) would be read from both of us for direct comparison of our brains' effort during

BRAIN WAVES

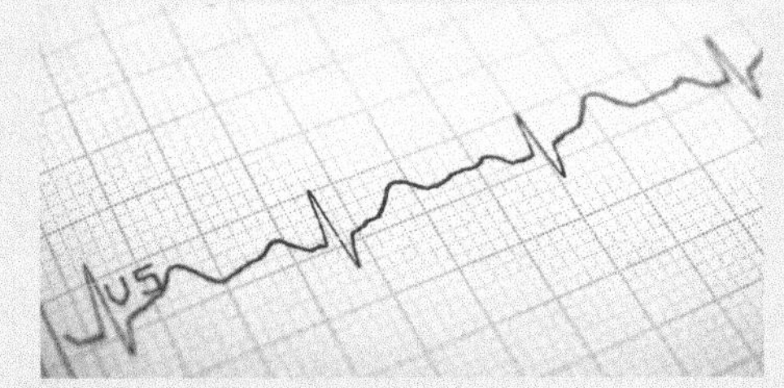

An EEG, short for electroencephalogram, is a method for eavesdropping on the overall electrical activity that arises from the activity of neurons. Small electrodes placed on the surface of the scalp pick up on "brain waves," the colloquial term for the averaged electrical signals produced by the underlying detailed neural chatter.

German physiologist and psychiatrist Hans Berger recorded the first human EEG in 1924, and researchers in the 1930s and 1940s identified several different types of brain waves: Delta waves (below 4 Hz) occur during sleep; Theta waves (4–7 Hz) are associated with sleep, deep relaxation, and visualization; Alpha waves (8–13 Hz) occur when we are relaxed and calm; Beta waves (13–38 Hz) are seen when we are actively thinking and problem solving. Other ranges of brain waves have been identified as important since then, including Gamma waves (39–100 Hz) which are involved in concentrated mental activity, such as reasoning and planning.

Our overall brain activity is a mix of all these different frequencies, but depending on what we're doing we'll exhibit some more than others.

the task. With both of us rigged up, we now had a crude window into the world inside our skulls.

Austin walked me through the steps of his routine. So as not to get smoked too badly by a ten-year-old, I practiced over and over for about twenty minutes before the official challenge began.

My efforts made no difference in the end. Austin beat me. I wasn't even an eighth of the way through the routine when he slammed the cups victoriously into their final configuration.

The defeat was not unexpected, but what did the EEG reveal? If Austin runs this routine eight times as quickly, it seems a reasonable assumption that it would cost him that much more energy. But that assumption overlooks a basic rule about how brains take on new skills. As it turns out, the EEG result showed that my brain, not Austin's, was the one working overtime, burning an enormous amount of energy to run this complex new task. My EEG showed high activity in the Beta wave frequency band, which is associated with extensive problem solving. Austin, on the other hand, had high activity in the Alpha wave band, a state associated with the brain at rest. Despite the speed and complexity of his actions, Austin's brain was serene.

Austin's talent and speed is the end result of physical changes in his brain. During his years of practice, specific patterns of physical connections have formed. He has carved the skill of cup stacking into the structure of his neurons. As a consequence, Austin now expends much less energy to stack cups. My brain, in contrast, is attacking the problem by conscious deliberation. I'm using general-purpose cognitive software; he's transferred the skill into specialized cognitive hardware.

When we practice new skills, they become physically hardwired, sinking below the level of consciousness. Some people are tempted to call this muscle memory, but in fact the skills are not stored in the muscles: instead, a routine like cup stacking is orchestrated across the thick jungles of connections in Austin's brain.

The detailed structure of the networks in Austin's brain has changed with his years of cup-stacking practice. A procedural memory is a long-term memory that represents how to do things automatically, like riding a bicycle or tying shoelaces. For Austin, cup stacking has become a procedural memory that is written into the microscopic hardware of his brain, making his actions both rapid and energy-efficient. Through practice, repeated signals have been passed along neural networks, strengthening synapses and thereby burning the skill into the circuitry. In fact, Austin's brain has developed such expertise that he can run flawlessly through the cup-stacking routine while wearing a blindfold.

In my case, as I learn to stack cups, my brain is enlisting slow, energy-hungry areas like the prefrontal cortex, parietal cortex and cerebellum—all of which are no longer needed for Austin to run the routine. In the early days of learning a new motor skill, the cerebellum plays a particularly important role, coordinating the flow of movements required for accuracy and perfect timing.

As a skill becomes hardwired, it sinks below the level of conscious control. At that point, we can perform a task automatically and without thinking about it—that is, without conscious awareness. In some cases, a skill is so hardwired that the circuitry underlying it is found below the brain, in the spinal cord. This has been observed in cats who have had much of their brain removed, and yet can still walk normally on a treadmill: the complex programs involved in gait are stored at a low level of the nervous system.

RUNNING ON AUTOPILOT

Throughout our lives, our brains rewrite themselves to build dedicated circuitry for the missions we practice—whether that's walking, surfing, juggling, swimming, or driving. This ability to burn programs into the structure of the brain is one of its most powerful tricks. It can solve the problem of complex movement using such little energy by wiring dedicated circuitry into the hardware. Once etched into the circuitry of the brain these skills can be run without thinking—without conscious effort—and this frees up resources, allowing the conscious me to attend to, and absorb, other tasks.

There is a consequence to this automization: new skills sink below the reach of conscious access. You

SYNAPSES AND LEARNING

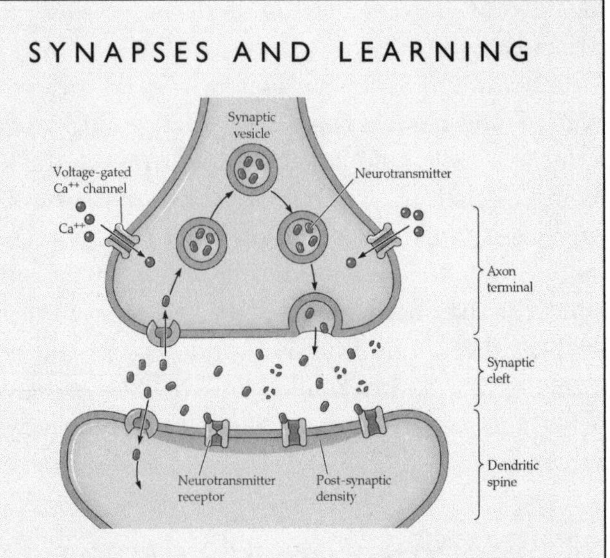

The connections between neurons are called synapses. These connections are where chemicals called neurotransmitters carry signals between neurons. But synaptic connections are not all of the same strength: depending on their history of activity, they can become stronger or weaker. As synapses change their potency, information flows through the network differently. If a connection gets weak enough, it withers and goes away. If it gets strengthened, it can sprout new connections. Some of this reconfiguring is guided by reward systems, which globally broadcast a neurotransmitter called dopamine when something has gone well. Austin's brain networks have been reshaped—very slowly, very subtly—by the success or failure of each attempted move, over hundreds of hours of practice.

lose access to the sophisticated programs running under the hood, so you don't know precisely how you do what you do. When you walk up a flight of stairs while having a conversation, you have no idea how you calculate the dozens of micro-corrections of your body's balance and how your tongue dynamically whips around to produce the right sounds for your language. These are difficult tasks that you couldn't always do. But because your actions become automatic and unconscious, this begets your capacity to run on autopilot. We all know the feeling of driving home along your daily route and suddenly realizing you've arrived with no real memory of the drive. The skills involved in driving have become so automatized that you can run the routines unconsciously. The conscious you—the part that flickered to life when you woke up in the morning—is no longer the driver, but at best a passenger along for the ride.

There's an interesting upshot to automatized skills: attempts to consciously interfere with them typically worsen their performance. Learned proficiencies—even very complex ones—are best left to their own devices.

Consider rock climber Dean Potter: until his recent death, he scaled cliffs without a rope and without safety equipment. From the age of twelve, Dean dedicated his life to climbing. Years of practice hardwired great precision and skill into his brain. To achieve his rock-climbing prowess, Dean relied on these overtrained circuits to do their work, unimpeded by conscious deliberation. He gave over complete control to his unconscious. He climbed in a brain state often referred to as "flow," a state in which extreme athletes commonly enjoy the far limits of their capacities. Like many athletes, Dean found his way into the flow state by putting himself in life-threatening danger. In that state, he experienced no meddling from his inner voice, and he could rely completely on the climbing abilities carved into his hardware over years of dedicated training.

Like cup-stacking champion Austin Naber, the brain waves of an athlete in flow are not crazed by the chatter of conscious deliberation (Do I look good? Should I have said such-and-such? Did I lock the door behind me?). During flow, the brain enters a state of hypofrontality, meaning that parts of the prefrontal cortex temporarily become less active. These are areas involved in abstract thinking, planning into the future, and concentrating on one's sense of self. Dialing down these background operations is the key move that allows a person to hang halfway up a rock face; feats like Dean's can only be done without the distraction of internal prattle.

It's often the case that consciousness is best left at the sidelines—and for some types of tasks, there's really no choice, because the unconscious brain can perform at speeds that the conscious mind is too slow to keep up with. Take the game of baseball, in which a fastball can travel from the pitcher's mound to the home plate at one hundred miles an hour. In order to make contact with the ball, the brain has only about four-tenths of a second to react. In that time it has to process and orchestrate an intricate sequence of movements to hit the ball. Batters connect with balls all the time, but they're not doing it consciously: the ball simply travels too quickly for the athlete to be consciously aware of its position, and the hit is over before the batter can register what happened. Not only has consciousness been left on the sidelines, it's also been left in the dust.

THE DEEP CAVERNS OF THE UNCONSCIOUS

The reach of the unconscious mind extends beyond control of our bodies. It shapes our lives in more profound ways. The next time you're in a conversation, notice the way words spill out of your mouth more quickly than you could possibly consciously control every word you say. Your brain is working behind the scenes, crafting and producing language, conjugations, and complex thoughts for you. (For comparison, compare your speed when speaking a foreign language that you're just learning!)

The same behind-the-scenes work is true of ideas. We take conscious credit for all our ideas, as though we've done the hard work in generating them. But in fact, your unconscious brain has been working on those ideas—consolidating memories, trying out new combinations, evaluating the consequences—for hours or months before the idea rises to your awareness and you declare, "I just thought of something!"

The man who first began to illuminate the hidden depths of the unconscious was one of the most influential scientists of the twentieth century. Sigmund Freud entered medical school in Vienna in 1873, and specialized in neurology. When he opened his private practice for the treatment of psychological disorders, he realized that often his patients had no conscious knowledge of what was driving their behavior. Freud's insight was that much of their behavior was a product of unseen mental processes. This simple idea transformed psychiatry, ushering in a new way of understanding human drives and emotions.

Before Freud, aberrant mental processes went unexplained or were described in terms of demonic possession, weak will, and so on. Freud insisted on seeking the cause in the physical brain.

He had patients lie down on a couch in his office so that they didn't have to look directly at him, and then he would get them to talk. In an era before brain scans, this was the best window into the world of the unconscious brain. His method was to gather information in patterns of behavior, in the content of dreams, in slips of the tongue, in mistakes of the pen. He observed like a detective, seeking clues to the unconscious neural machinery to which patients had no direct access.

He became convinced that the conscious mind is the tip of the iceberg of our mental processes, while the much larger part of what drives our thoughts and behaviors lies hidden from view.

Freud's speculation turned out to be correct, and one consequence is that we don't typically know the roots of our own choices. Our brains constantly pull information from the environment and use it to steer our behavior, but often the influences around us are not recognized. Take an effect called "priming," in which one thing influences the perception of something else. For example, if you're holding a warm drink you'll describe your relationship with a family member more favorably; when you're holding a cold drink, you'll express a slightly poorer opinion of the relationship. Why does this happen? Because the brain mechanisms for judging intrapersonal warmth overlap with the mechanisms for judging physical warmth, and so one influences the other. The upshot is that your opinion about something as fundamental as your relationship with your mother can be manipulated by whether you take your tea hot or iced. Similarly, when you are in a foul-smelling environment, you'll make harsher moral decisions—for example, you're more likely to judge that someone else's uncommon actions are immoral. In another study, it was shown that if you sit in a hard chair you'll be a more hard-line negotiator in a business transaction; in a soft chair you'll yield more.

Take as another example the unconscious influence of "implicit egotism," which describes our attraction to things that remind us of ourselves. When social psychologist Brett Pelham and his team analyzed the records of graduates from dental and law schools, they found a statistical overrepresentation of dentists named Dennis or Denise, and of lawyers named Laura or Laurence. They also found that owners of roofing companies are more likely to have a first name beginning with R while hardware store owners are more likely to have a first name beginning with H. But is our career choice the only place where we make these decisions? It turns out that our love lives may be heavily influenced by these similarities too. When psychologist John Jones and his colleagues looked at the marriage registers in Georgia and Florida they discovered that more married couples than expected shared the same first initial. This means that Jenny is more likely to marry Joel, Alex marry Amy, and Donny marry Daisy. These kinds of unconscious effects are small but verifiable.

Here's the critical point: if you were to ask any of these Dennises or Lauras or Jennys why they chose their profession or their mate, they would have a conscious narrative to give you. But that narrative wouldn't include the long reach of their unconscious on some of their most important life choices.

Take another experiment designed by psychologist Eckhard Hess in 1965. Men were asked to look at photographs of women's faces and make judgments about them. How attractive were they, on a scale from one to ten? Were they happy or sad? Mean or kind? Friendly or unfriendly? Unbeknownst to the participants, the photographs had been manipulated. In half of the photographs, the women's pupils had been artificially dilated.

NUDGING THE UNCONSCIOUS

In their book *Nudge*, Richard Thaler and Cass Sunstein laid out an approach to improving "decisions about health, wealth, and happiness" by playing to the brain's unconscious networks. A small nudge in our environment can change our behavior and decision making for the better, without us being aware of it. Placing fruit at eye level in supermarkets nudges people to make healthier food choices. Pasting a picture of a housefly in urinals at airports nudges men to aim better. Automatically opting employees into retirement plans (with the freedom to opt out if they'd like to) leads to better saving practices. This view of governance is called soft paternalism, and Thaler and Sunstein believe that gently guiding the unconscious brain has a far more powerful influence on our decision making than outright enforcement ever can.

The men found the women with dilated eyes to be more attractive. None of the men explicitly noted anything about women's pupil sizes—and presumably none of the men knew that dilated eyes are a biological sign of female arousal. But their brains knew it. And the men were unconsciously steered toward the women with the dilated eyes, finding them to be more beautiful, happier, kinder, and more friendly.

Really, this is how love often goes. You find yourself more attracted to some people over others, and it's generally not possible to put your finger on precisely why. Presumably there is a why; you just don't have access to it.

In another experiment, evolutionary psychologist Geoffrey Miller quantified how sexually attractive a woman is to a man by recording the earnings of lap dancers in a strip club. And he tracked how this changed over their monthly menstruation cycle. As it turned out, men gave twice as much in tips when the dancer was ovulating (fertile) as when she was menstruating (not fertile). But the strange part is that the men weren't consciously aware of the biological changes that attend the monthly cycle—that when she is ovulating, a surge of the hormone estrogen changes her appearance subtly, making her features more symmetrical, her skin softer, and her waist narrower. But they detected these fertility cues nonetheless, under the radar of awareness.

These kinds of experiments reveal something fundamental about how brains operate. The job of this organ is to gather information about the world and steer your behavior appropriately. It doesn't matter if your conscious awareness is involved or not. And most of the time it's not. Most of the time you are not aware of the decisions being made on your behalf.

WHY ARE WE CONSCIOUS?

So why aren't we just unconscious beings? Why aren't we all wandering around like mindless zombies? Why did evolution build a brain that's conscious? To answer this, imagine walking along a local street, minding your own business. All of a sudden something catches your eye: someone ahead of you is dressed up in a giant bee costume, holding a briefcase. If you were to watch the human bee, you'd notice how people who catch a glimpse of him react: they break out of their automated routines and stare.

Consciousness gets involved when the unexpected happens, when we need to work out what to do next. Although the brain tries to tick along as long as possible on autopilot, it's not always possible in a world that throws curveballs.

But consciousness isn't just about reacting to surprises. It also plays a vital role in settling conflict within the brain. Billions of neurons participate in tasks ranging from breathing to moving through your bedroom to getting food into your mouth to mastering a sport. These tasks are each underpinned by vast networks in the machinery of the brain. But what happens if there's a conflict? Say you find yourself reaching for an ice cream sundae, but you know that you'll regret having eaten it. In a situation like that, a decision has to be made. A decision that works out what's best for the organism—you—and your long-term goals. Consciousness is the system that has this unique vantage point, one that no other subsystem of the brain has. And for this reason, it can play the role of arbiter of the billions of interacting elements, subsystems and burned-in processes. It can make plans and set goals for the system as a whole.

I think of consciousness as the CEO of a large sprawling corporation, with many thousands of subdivisions and departments all collaborating and interacting and competing in different ways. Small companies don't need a CEO—but when an organization reaches sufficient size and complexity, it needs a CEO to stay above the daily details and to craft the long view of the company.

Although the CEO has access to very few details of the day-to-day running of the company, he or she always has the long view of the company in mind. A CEO is a company's most abstract view of itself. In terms of the brain, consciousness is a way for billions of cells to see themselves as a unified whole, a way for a complex system to hold up a mirror to itself.

WHEN CONSCIOUSNESS GOES MISSING

What if consciousness doesn't kick in and we are lost in autopilot for too long?

Ken Parks, aged twenty-three, found out on May 23, 1987, when he fell asleep at home while watching TV. At the time, he lived with his five-month-old daughter and his wife, and was going through financial difficulties, marital problems and a gambling addiction. He had planned to discuss his problems with his in-laws the following day. His mother-in-law described him as a "gentle giant" and he got along well with both of his wife's parents. At some point during the night, he got up, drove twenty-three kilometers to his in-laws' house, strangled his father-in-law, and stabbed his mother-in-law to death. He then drove to the nearest police station, and said to the officer, "I think I just killed someone."

He had no memory of what had happened. It seemed somehow that his conscious mind was absent during this horrific episode. What had gone wrong with Ken's brain? Parks's lawyer, Marlys Edwardh, assembled a team of experts to help figure out this mystery. They soon began to suspect the events might be connected to Ken's sleep. While Ken was in prison, his lawyer called in sleep expert Roger Broughton, who measured Ken's EEG signals while he slept at night. The recorded output was consistent with that of a sleepwalker.

As the team investigated further, they found sleep disorders throughout Ken's extended family. With no motive, no way to fake his sleep results, and such extensive family history, Ken was found not guilty of homicide, and he was released.

SO WHO IS IN CONTROL?

All this might leave you wondering what control the conscious mind really has. Is it possible that we are living our lives like puppets at the mercy of a system that is pulling our strings and determining what we do next? There are some who believe this is the case and that our conscious minds have no control over what we do.

Let's dig into this question via a simple example. You drive up to a fork in the road where you can either turn left or right. There is no obligation for you to turn one way or the other, but today, at this moment, you feel like you want to turn right. So you turn right. But why did you turn right, and not left? Because you felt like it? Or because inaccessible mechanisms in your brain decided it for you? Consider this: the neural signals that move your arms to turn the steering wheel come from your motor cortex, but those signals don't originate there. They're driven by other regions of the frontal lobe, which are in turn driven by many other parts of the brain, and so on in a complex linkage that crisscrosses the brain's entire network. There is never a time zero when you decide to do something, because every neuron in the brain is driven by other neurons; there seems to be no part of the system that acts independently rather than reacts dependably. Your decision to turn right—or left—is a decision that reaches back in time: seconds, minutes, days, a lifetime. Even when decisions seem spontaneous, they don't exist in isolation.

So when you roll up to that fork in the road carrying your lifetime's history with you, who exactly is responsible for the decision? These considerations lead to the deep question of free will. If we rewound history one hundred times, would you always do the same thing?

THE FEELING OF FREE WILL

We feel like we have autonomy—that is, we make our choices freely. But under some circumstances it's possible to demonstrate that this feeling of autonomy can be illusory. In one experiment, Professor Alvaro Pascual-Leone at Harvard invited participants to his lab for a simple experiment.

The participants sat in front of a computer screen with both hands outstretched. When the screen turned red, they would make an internal choice about which hand they were going to move—but they wouldn't actually move. Then the light turned yellow, and when it finally turned green the person activated their prechosen move, lifting either their right or left hand.

Then the experimenters introduced a twist. They used transcranial magnetic stimulation (TMS), which discharges a magnetic pulse and excites the area of the brain underneath, to stimulate the motor cortex and initiate movement in either the left or right hand. Now, during the yellow light, they gave the TMS pulse (or, in the control condition, just the sound of the pulse).

The TMS intervention made subjects favor one hand over another—for example, stimulation over the left motor cortex made participants more likely to lift their right hand. But the interesting part was that subjects reported the feeling of having wanted to move the hand that was being manipulated by TMS. In other words, they might internally choose to move their left hand during the red light, but then, after stimulation during the yellow light, they might feel that they really had wanted to move their right hand all along. Although the TMS was initiating the movement in their hand, many of the participants felt as if they had made decisions of their own free will. Pascual-Leone reports that participants often said they had meant to switch their choice. Whatever the activity in their brain was up to, they took credit for it as though it were freely chosen. The conscious mind excels at telling itself the narrative of being in control.

Experiments like these expose the problematic nature of trusting our intuitions about the freedom of our choices. At the moment, neuroscience doesn't have the perfect experiments to entirely rule free will out; it's a complex topic, and one that our science may simply be too young to address thoroughly. But let's entertain for a moment the prospect that there really is no free will; when you arrive at that fork in the road, your choice is predetermined. On the face of it, a life that's predictable doesn't sound like a life worth living.

The good news is that the brain's immense complexity means that in actuality, nothing is predictable. Imagine a tank with rows of Ping-Pong balls along the bottom—each one delicately poised on its own mousetrap, sprung and ready. If you were to drop in one more Ping-Pong ball from the top, it should be relatively straightforward to mathematically predict where it will land. But as soon as that ball hits the bottom, it sets off an unpredictable chain reaction. It triggers other balls to be flung from their mousetraps, and those trigger yet other balls, and the situation quickly explodes in complexity. Any error in the initial prediction, no matter how small, becomes magnified as balls collide and bounce off the sides and land on other balls. Soon it's utterly impossible to make any kind of forecast about where the balls will be.

Our brains are like this Ping-Pong ball tank, but massively more complex. You might be able to fit a few hundred Ping-Pong balls in a tank, but your skull houses trillions of times more interactions than the tank, and it goes on bouncing throughout every second of your lifetime. And from those innumerable exchanges of energy, your thoughts, feelings, and decisions emerge.

And this is only the beginning of the unpredictability. Each individual brain is embedded in a world of other brains. Across the space of a dinner table, or the length of a lecture hall, or the reach of the Internet, all the human neurons on the planet are influencing one other, creating a system of unimaginable complexity. This means that even though neurons follow straightforward physical rules, in practice it will always be impossible to predict exactly what any individual will do next.

This titanic complexity leaves us with just enough insight to understand a simple fact: our lives are steered by forces far beyond our capacity for awareness or control.

4

HOW DO I DECIDE?

Should I eat the ice cream or not? Do I answer this email now or later? Which shoes? Our days are assembled from thousands of small decisions: what to do, which way to go, how to respond, whether to partake. Early theories of decision making assumed that humans are rational actors, tallying the pros and cons of our options to come to an optimal decision. But scientific observations of human decision making don't bear that out. Brains are composed of multiple, competing networks, each of which has its own goals and desires. When deciding whether or not to gobble down the ice cream, some networks in your brain want the sugar; other networks vote against it based on long-term considerations of vanity; other networks suggest that perhaps you could eat the ice cream if you promise yourself you'll go to the gym tomorrow. Your brain is like a neural parliament, composed of rival political parties which fight it out to steer the ship of state. Sometimes you decide selfishly, sometimes generously, sometimes impulsively, and sometimes with the long view in mind. We are complex creatures because we are composed of many drives, all of which want to be in control.

THE SOUND OF A DECISION

On the operating table, a patient named Jim is undergoing brain surgery to stop tremors of his hand. Long, thin wires called electrodes have been lowered into Jim's brain by the neurosurgeon. By applying a small electric current through the wires, the patterns of activity in Jim's neurons can be adjusted to reduce his tremors.

The electrodes create a special opportunity to eavesdrop on the activity of single neurons. Neurons talk with one another via electrical spikes called action potentials, but these signals are invisibly tiny, so surgeons and researchers often pass the tiny electrical signals through an audio speaker. That way, a miniscule change in voltage (a tenth of a volt that lasts a thousandth of a second) is turned into an audible pop!

As the electrode is lowered through different regions of the brain, the activity patterns of those regions can be recognized by the trained ear. Some locations are characterized by *pop!pop!pop!* while others sound quite different: *pop!....poppop!...pop!* It's like suddenly dropping in on the conversation of a few people somewhere randomly on the globe: because the people you land upon will have specific jobs in diverse cultures, they'll all have very different conversations going on.

I'm in the operating room as a researcher: while my colleague performs the surgery, my goal is to better understand how the brain makes decisions. To that end, I ask Jim to perform different tasks—like speaking, reading, looking, deciding—to determine what's correlated with the activity of his neurons. Because the brain has no pain receptors, a patient can be awake during a surgery. I ask Jim to look at a simple picture while we're recording.

What happens in your brain when you see the old woman? What changes when you see the young lady?

In the figure, you may see a young lady with a bonnet looking away. Now try to find another way of interpreting the same image: an old woman looking down and to the left. This picture can be seen in one of two ways (this is known as perceptual bi-stability): the lines on the page are consistent with two very different interpretations. When you stare at the figure, you'll see one version, and then eventually the other, and then the first again, and so on. Here's the important part: nothing on the physical page changes—so whenever Jim reports that the image has flipped, it has to be because of something that changed inside his brain.

The moment he sees the young lady, or the old woman, his brain has made a decision. A decision doesn't have to be conscious; in this case, it's a perceptual decision by Jim's visual system, and the mechanics of the switchover are hidden completely under the hood. In theory, a brain should be able to see both the young lady and the old lady at the same time—but in reality a brain doesn't do that. Reflexively, it takes something ambiguous and makes a choice. Eventually, it remakes the choice, and it might switch back and forth over and over. But our brains are always crushing ambiguity into choices.

So when Jim's brain lands on an interpretation of the young lady—or the old woman—we can listen to the responses from a small number of neurons. Some leap into a higher rate of activity (poppop!.pop!..pop!), while other neurons slow down (pop!....pop!..pop!.... pop!). It's not always about speeding up and slowing down: sometimes neurons change their pattern of activity in more subtle ways, becoming synchronized or desynchronized with other neurons even while maintaining their original pace.

The neurons we happen to be spying on are not, by themselves, responsible for the perceptual change—

instead, they operate in concert with billions of other neurons, so the changes we can witness are just the reflection of a changing pattern taking hold across large sweeps of brain territory. When one pattern wins out over the other in Jim's brain, a decision has been landed upon.

Your brain makes thousands of decisions every day of your life, dictating your experience of the world. From the decision of what to wear, whom to call, how to interpret an offhand comment, whether to reply to an e-mail, when to leave—decisions underlie our every action and thought. Who you are emerges from the brain-wide battles for dominance that rage in your skull every moment of your life.

Listening to Jim's neural activity—pop!pop!pop!—it's impossible not to be awed. After all, this is what every decision in the history of our species sounded like. Every marriage proposal, every declaration of war, every leap of the imagination, every mission launched into the unknown, every act of kindness, every lie, every euphoric breakthrough, every decisive moment. It all happened right here, in the darkness of the skull, emerging from patterns of activity in networks of biological cells.

THE BRAIN IS A MACHINE BUILT FROM CONFLICT

Let's take a closer look at what's happening behind the scenes during a decision. Imagine you're making a simple choice, standing in the frozen yogurt store, trying to decide between two flavors you like equally. Say these are mint and lemon. From the outside, it doesn't look like you're doing much: you're simply stuck there, looking back and forth between the two options. But inside your brain, a simple choice like this unleashes a hurricane of activity.

By itself, a single neuron has no meaningful influence. But each neuron is connected to thousands of others, and they in turn connect to thousands of others, and so on in a massive, loopy, intertwining network. They're all releasing chemicals that excite or depress each other.

Within this web, a particular constellation of neurons represents mint. This pattern is formed from neurons that mutually excite each other. They're not necessarily next to one another; rather, they might span distant brain regions involved in smell, taste, vision, and your unique history of memories involving mint. Each of these neurons, by itself, has little to do with mint—in fact, each neuron plays many roles, at different times, in ever-shifting coalitions. But when these neurons all become active collectively, in this particular arrangement . . . that's mint to your brain. As you're standing in front of the yogurt selection, this federation of neurons eagerly communicates with one another like dispersed individuals linking online.

These neurons aren't acting alone in their electioneering. At the same time, the competing possibility—lemon—is represented by its own neural party. Each coalition—mint and lemon—tries to gain the upper hand by intensifying its own activity and suppressing the other's. They fight it out until one triumphs in the winner-take-all competition. The winning network defines what you do next.

Unlike computers, the brain runs on conflict between different possibilities, all of which try to out-compete the others. And there are always multiple options. Even after you've selected mint or lemon, you find yourself in a new conflict: should you eat the whole thing? Part of you wants the delicious energy source, and at the same time part of you knows it's sugary, and perhaps you should be jogging instead. Whether you polish off the whole container is simply a matter of the way the infighting goes.

As a result of ongoing conflicts in the brain, we can argue with ourselves, curse at ourselves, cajole ourselves. But who exactly is talking with whom? It's all you—but it's different parts of you.

To tease apart some of the major competing systems in the brain, consider a thought experiment known as the trolley dilemma. A trolley is barreling down a train track, out of control. Four workers are making repairs farther down the track, and you, a bystander, quickly realize that they will all be killed by the runaway trolley. Then you notice that there's a lever nearby that can divert the trolley onto another track. But hang on! You see that there's one worker on that track. So if you pull the lever, one worker will be killed; if you don't, four will be killed. Do you pull the lever?

THE SPLIT BRAIN: UNMASKING THE CONFLICT

Under special circumstances it becomes particularly easy to witness internal conflict between the different parts of the brain. As a treatment for certain forms of epilepsy, some patients undergo "split-brain" surgery, in which the brain's two hemispheres are disconnected from each other. Normally the two hemispheres are connected by a superhighway of nerves called the corpus callosum, and this allows the right and left halves to coordinate and work in concert. If you're feeling chilly, both of your hands cooperate: one holds your jacket hem while the other tugs up the zipper.

But when the corpus callosum is severed, a remarkable and haunting clinical condition can emerge: alien hand syndrome. The two hands can act with totally different intentions: the patient begins to zip up a jacket with one hand, and the other hand (the "alien" hand) suddenly grabs the zipper and pulls it back down. Or the patient might reach for a biscuit with one hand, and their other hand leaps into action to slap the first hand into failure. The normal conflict running in the brain is revealed as the two hemispheres act independently of each other.

Alien hand syndrome normally fades in the weeks after surgery, as the two halves of the brain take advantage of remaining connections to begin coordinating again. But it serves as a clear demonstration that even when we think we're being single-minded, our actions are the product of immense battles that continually rise and fall in the darkness of the cranium.

The trolley dilemma. When people are asked what they would do in this scenario, almost everyone pulls the lever. After all, it's far better that only one person is killed rather than four, right?

Now consider a slightly different, second scenario. The situation begins with the same premise: a trolley is barreling down the tracks, out of control, and four workers are going to be killed. But this time you're standing on the deck of a water tower overlooking the

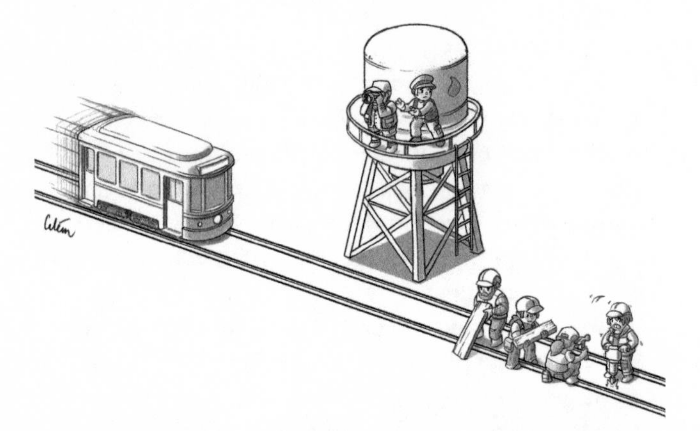

The trolley dilemma, scenario 2. In this situation, almost no one is willing to push the man. Why not? When asked, they give answers like "that would be murder" and "that would just be wrong."

tracks, and you notice there's a large man standing up there with you, gazing out into the distance. You realize that if you push him off, he'll land right on the track—and his body weight will be sufficient to stop the trolley and save the four workers.

Do you push him off?

But wait. Aren't you being asked to consider the same equation in both cases? Trading one life for four? Why do the results come out so differently in the second scenario? Ethicists have addressed this problem from many angles, but neuroimaging has been able to provide a fairly straightforward answer. To the brain, the first scenario is just a math problem. The dilemma activates regions involved in solving logical problems.

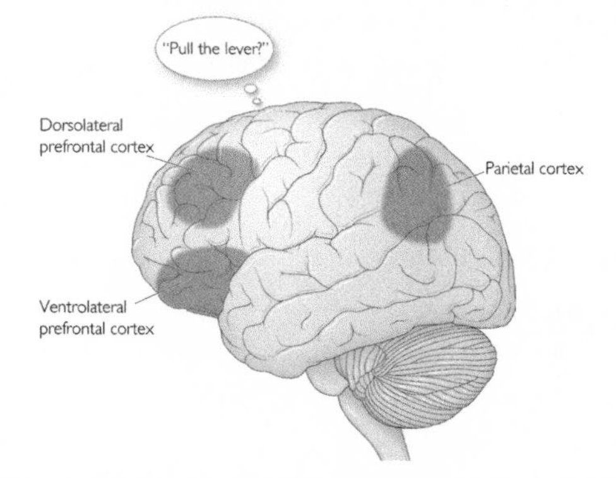

Several regions of the brain are more invested in logical problem solving.

In the second scenario, you have to physically interact with the man and push him to his death. That recruits additional networks into the decision: brain regions involved in emotion.

In the second scenario, we're caught in a conflict between two systems that have different opinions. Our rational networks tell us that one death is better than four, but our emotional networks trigger a gut feeling that murdering the bystander is wrong. You're caught between competing drives, with the result that your decision is likely to change entirely from the first scenario.

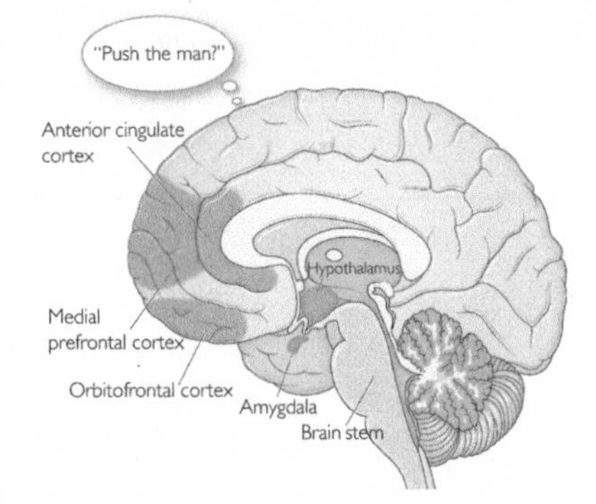

When considering pushing an innocent man to his death, networks involved in emotions become more involved in the decision making—and that can flip the outcome.

The trolley dilemma sheds light on real-world situations. Consider modern warfare, which has become more like pulling the lever than pushing the man off the tower. When a person hits the button to launch a long-range missile, it involves only the networks involved in solving logical problems. Operating a drone can become like a video game; cyberattacks wreak consequences at a distance. The rational networks are at work here, but not necessarily the emotional networks. The detached nature of distance warfare reduces internal conflict, making it easier to wage.

One pundit suggested that the button to launch nuclear missiles should be implanted in the chest of the President's best friend. That way, if he chose to launch nukes, he'd have to inflict physical violence on his friend, tearing him open. That consideration would recruit emotional networks into the decision. When making life-and-death decisions, unchecked reason can be dangerous; our emotions are a powerful and often insightful constituency, and we'd be remiss to exclude them from the parliamentary voting. The world would not be better if we all behaved like robots.

Although the neuroscience is new, this intuition has a long history. The ancient Greeks suggested that we should think of our lives like chariots. We are charioteers trying to hold two horses: the white horse of reason and the black horse of passion. Each horse pulls off-center, in opposite directions. Your job is to keep control of both horses, navigating down the middle of the road.

Indeed, in typical neuroscientific fashion, we can unmask the importance of emotions by seeing what happens when someone loses the capacity to include them in decision making.

STATES OF THE BODY HELP YOU DECIDE

Emotions do more than add richness to our lives—they're also the secret behind how we navigate what to do next at every moment. This is illustrated by looking at the situation of Tammy Myers, a former engineer who got into a motorcycle accident. The consequence was damage to her orbitofrontal cortex, the region just above the sockets of the eyes. This brain region is critical for integrating signals streaming in from her body—signals that tell the rest of the brain what state her body is in: hungry, nervous, excited, embarrassed, thirsty, joyful.

Tammy doesn't look like someone who has suffered a traumatic brain injury. But if you were to spend even five minutes with her, you would detect that there's a problem with her ability to handle life's daily decisions. Although she can describe all the pros and cons of a choice in front of her, even the simplest situations leave her mired in indecision. Because she can no longer read her body's emotional summaries, decisions become incredibly difficult for her. Now, no choice is tangibly different from another. Without decision making, little gets done; Tammy reports she often spends all day on the sofa.

Tammy's brain injury tells us something crucial about decision making. It's easy to think about the brain commanding the body from on high—but in fact the brain is in constant feedback with the body. The physical signals from the body give a quick summary of what's going on and what to do about it. To land on a choice, the body and the brain have to be in close communication.

Consider this situation: you want to pass a mis-delivered package over to your next-door neighbors. But as you approach the gate to their yard, their dog growls and bares its teeth. Do you open the gate and press on to the front door? Your knowledge of the statistics of dog attacks isn't the deciding factor here—instead, the dog's threatening posture triggers a set of physiological responses in your body: an increased heart rate, a tightening in the gut, a tensing of the muscles, pupil dilation, changes in blood hormones, opening of sweat glands, and so on. These responses are automatic and unconscious.

In this moment, standing there with your hand on the gate latch, there are many external details you could assess (for example, the color of the dog's collar)—but what your brain really needs to know right now is whether you should face the dog or deliver the package another way. The state of your body helps you in this task: it serves as a summary of the situation. Your physiological signature can be thought of as a low-resolution headline: "this is bad" or "this is no problem." And that helps your brain decide what to do next.

Every day we read the states of our bodies like this. In most situations, physiological signals are more subtle, and so we tend to be unaware of them. However, those signals are crucial to steering the decisions we have to make. Consider being in a supermarket: this is the kind of place which leaves Tammy paralyzed with indecision. Which apples? Which bread? Which ice cream? Thousands of choices bear down upon shoppers, with the end result that we spend hundreds of hours of our lives standing in the aisles, trying to make our neural networks commit to one decision over another. Although we don't commonly realize it, our body helps us to navigate this boggling complexity.

Take the choice of which kind of soup to buy. There's too much data here for you to grapple with: calories, price, salt content, taste, packaging, and so on. If you were a robot, you'd be stuck here all day trying to make a decision, with no obvious way to trade off which details matter more. To land on a choice, you need a summary of some sort. And that's what the feedback from your body is able to give you. Thinking about your budget might make your palms sweat, or you might salivate thinking about the last time you consumed the chicken noodle soup, or noting the excessive creaminess of the other soup might put a cramp in your intestines. You simulate your experience with one soup, and then the other. Your bodily experience helps your brain to quickly place a value on soup A, and another on soup B, allowing you to tip the balance in one direction or the other. You don't just extract the data from the soup cans, you feel the data. These emotional signatures are more subtle than the ones related to facing down a barking dog, but the idea is the same: each choice is marked by a bodily signature. And that helps you to decide.

Earlier, when you were deciding between the mint and lemon yogurt, there was a battle between networks. The physiological states from your body were the key things that helped tip that battle, that allowed one network to win over another. Because of her brain damage, Tammy lacks the ability to integrate her bodily signals into her decision making. So she has no way to rapidly compare the overall value between options, no way to prioritize the dozens of details that she can articulate. That's why Tammy stays on the sofa much of the time: none of the choices in front of her carry any particular emotional value. There's no way to tip one network's campaign over any other. The debates in her neural parliament continue along in deadlock.

Because the conscious mind has low bandwidth, you don't typically have full access to the bodily signals that tip your decisions; most of the action in your body lives far below your awareness. Nonetheless, the signals can have far-reaching consequences on the type of person you believe you are. As one example, neuroscientist Read Montague has found a link between a person's politics and the character of their emotional responses. He puts participants in a brain scanner and measures their response to a series of images chosen to evoke a disgust response, from images of feces to dead bodies to insect-covered food. When they emerge from the scanner, they are asked if they would like to take part in another experiment; if they say "yes" they take ten minutes to answer a political ideology survey. They are asked questions about their feelings on gun control, abortion, premarital sex, and so on. Montague finds that the more disgusted a participant is by the images, the more politically conservative they are likely to be. The less disgusted, the more liberal. The correlation is so strong that a person's neural response to a single disgusting image predicts their score on the political ideology test with 95 percent accuracy. Political persuasion emerges at the intersection of the mental and the corporal.

TRAVELING TO THE FUTURE

Each decision involves our past experiences (stored in the states of our body) as well as the present situation (Do I have enough money to buy X instead of Y? Is option Z available?). But there's one more part to the story of decisions: predictions about the future.

Across the animal kingdom, every creature is wired to seek reward. What is a reward? At its essence, it's something that will move the body closer to its ideal set points. Water is a reward when your body is getting dehydrated; food is a reward when your energy stores are running down. Water and food are called primary rewards, which directly address biological needs. More generally, however, human behavior is steered by secondary rewards, which are things that predict primary rewards. For example, the sight of a metal rectangle wouldn't by itself do much for your brain, but because you've learned to recognize it as a water fountain, then the sight of it comes to be rewarding when you are thirsty. In the case of humans, we can find even very abstract concepts rewarding, such as the feeling that we are valued by our local community. And unlike animals, we can often put these rewards ahead of biological needs. As Read Montague points out, "sharks don't go on hunger strikes": the rest of the animal kingdom only chases its basic needs, while only humans regularly override those needs in deference to abstract ideals. So when we're faced with an array of possibilities, we integrate internal and external data to try to maximize reward, however it's defined to us as individuals.

The challenge with any reward, whether basic or abstract, is that choices typically don't yield their fruits right away. We almost always have to make decisions in which a chosen course of action returns reward at a later time. People go to school for years because they value the future concept of having a degree, they slave through employment they don't enjoy with the future hope of a promotion, and they push themselves through painful exercise with the goal of being fit.

To compare different options means assigning a value to each one in a common currency—that of anticipated reward—and then choosing the one with the highest value. Consider this scenario: I have a bit of free time and I'm trying to decide what to do. I need to get groceries, but I also know I need to get to a coffee shop and work on a grant for my lab, because a deadline is coming up. I also want to spend time with my son at the park. How I do arbitrate this menu of options?

It would be easy, of course, if I could directly compare these experiences by living each one, and then rewinding time, and finally choosing my path based on which outcome was the best. Alas, I cannot travel in time.

Or can I?

Time travel is something the human brain does relentlessly. When faced with a decision, our brains simulate different outcomes to generate a mock-up of what our future might be. Mentally, we can disconnect from the present moment and voyage to a world that doesn't yet exist.

Now, simulating a scenario in my mind is just the first step. To decide between the imagined scenarios, I try to estimate what the reward will be in each of those potential futures. When I simulate filling my pantry with the groceries, I feel a sense of relief at being organized and avoiding uncertainty. The grant carries different sorts of rewards: not only money for the laboratory, but more generally the kudos from my department chairman and a rewarding sense of accomplishment in my career. Imagining myself at the park with my son inspires joy, and a sense of reward in terms of family closeness. My final decision will be navigated by how each future stacks up against the others in the common currency of my reward systems. The choice isn't easy, because all these valuations are nuanced: the simulation of the grocery shopping is accompanied by feelings of tedium; the grant writing is attended by a sense of frustration; the park with guilt about not getting work done. Typically under the radar of awareness, my brain simulates all the options, one at a time, and does a gut check on each. That's how I decide.

How do I accurately simulate these futures? How

can I possibly predict what it will really be like to go down these paths? The answer is that I can't: there's no way to know that my predictions will be accurate. All my simulations are only based on my past experiences and my current models of how the world works. Like all animals in the animal kingdom, we can't just wander around hoping to randomly discover what results in future reward and what doesn't. Instead, the key business of brains is to predict. And to do this reasonably well, we need to continually learn about the world from our every experience. So in this case, I place a value on each of these options based on my past experiences. Using the Hollywood studios in our minds, we travel in time to our imagined futures to see how much value they'll have. And that's how I make my choices, comparing possible futures against one another. That's how I convert competing options into a common currency of future reward.

Think of my predicted reward value for each option like an internal appraisal that stores how good something will be. Because grocery shopping will supply me with food, let's say it's worth ten reward units. Grant writing is difficult but necessary to my career, so it weighs in at twenty-five reward units. I love spending time with my son, so going to the park is worth fifty reward units.

But there's an interesting twist here: the world is complicated, and so our internal appraisals are never written in permanent ink. Your valuation of everything around you is changeable, because quite often our predictions don't match what actually happens. The key to effective learning lies in tracking this *prediction error*: the difference between the expected outcome of a choice and the outcome that actually occurred.

In today's case, my brain has a prediction about how rewarding the park is going to be. If we run into friends there and it turns out even better than I thought, that raises the appraisal the next time I'm making such a decision. On the other hand, if the swings are broken and it rains, that lowers my appraisal the next time around.

How does this work? There's a tiny, ancient system in the brain whose mission is to keep updating your assessments of the world. This system is made of tiny groups of cells in your midbrain that speak in the language of a neurotransmitter called dopamine.

When there's a mismatch between your expectation and your reality, this midbrain dopamine system broadcasts a signal that reevaluates the price point. This signal tells the rest of the system whether things turned out better than expected (an increased burst of dopamine) or worse (a decrease in dopamine). That prediction error signal allows the rest of the brain to adjust its expectations to try to be closer to reality next time. The dopamine acts as an error corrector: a chemical appraiser that always works to make your appraisals as updated as they can be. That way, you can prioritize your decisions based on your optimized guesses about the future.

Fundamentally, the brain is tuned to detect unexpected outcomes—and this sensitivity is at the heart of animals' ability to adapt and learn. It's no surprise, then, that the brain architecture involved in learning from experience is consistent across species, from honeybees to humans. This suggests that brains discovered the basic principles of learning from reward long ago.

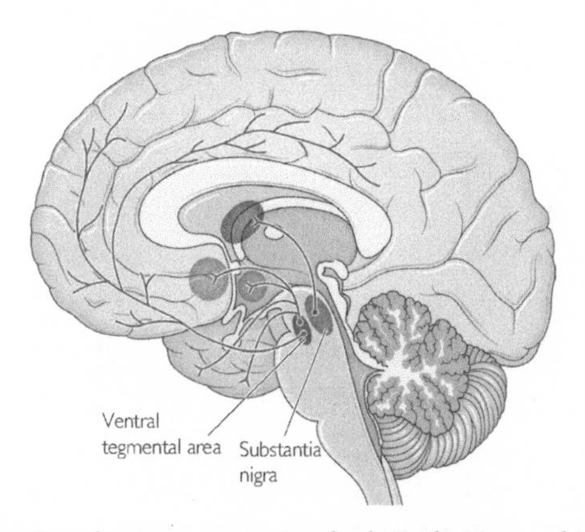

Dopamine-releasing neurons involved in decision making are concentrated into tiny regions of the brain called the ventral tegmental area and the substantia nigra. Despite their small sizes, they have a wide reach, broadcasting updates when the predicted value of a choice turns out to be too high or too low.

THE POWER OF NOW

So we've seen how values get attached to different options. But there's a twist that often gets in the way of good decision making: options right in front of us tend to be valued higher than those we merely simulate.

The thing that trips up good decision making about the future is the present.

In 2008, the US economy took a sharp downturn. At the heart of the trouble was the simple fact that many homeowners had overborrowed. They had taken out loans that offered wonderfully low interest rates for a period of a few years. The problem occurred at the end of the trial period, when the rates went up. At the higher rates, many homeowners found themselves unable to make the payments. Close to a million homes went into foreclosure, sending shock waves through the economy of the planet.

What did this disaster have to do with competing networks in the brain? These subprime loans allowed people to obtain a nice house now, with the high rates deferred until later. As such, the offer perfectly appealed to the neural networks that desire instant gratification—that is, those networks that want things now. Because the seduction of the immediate satisfaction pulls so strongly on our decision making, the housing bubble can be understood not simply as an economic phenomenon, but also as a neural one.

The pull of the now wasn't just about the people borrowing, of course, but also the lenders who were getting rich, right now, by offering loans that weren't going to get paid. They rebundled the loans and sold them off. Such practices are unethical, but the temptation proved too enticing to many thousands.

This now-versus-the-future battle doesn't just apply to housing bubbles, it cuts across every aspect of our lives. It's why car dealers want you to get in and test-drive the cars, why clothing stores want you to try on the clothes, why merchants want you to touch the merchandise. Your mental simulations can't live up to the experience of something right here, right now.

To the brain, the future can only ever be a pale shadow of the now. The power of now explains why people make decisions that feel good in the moment but have lousy consequences in the future: people who take a drink or a drug hit even though they know they shouldn't; athletes who take anabolic steroids even though it may shave years off their lives; married partners who give in to an available affair.

Can we do anything about the seduction of the now? Thanks to competing systems in the brain, we can. Consider this: we all know that it's difficult to do certain things, like go regularly to the gym. We want to be in shape, but when it comes down to it, there are usually things right in front of us that seem more enjoyable. The pull of what we're doing is stronger than the abstract notion of future fitness. So here's the solution: to make certain you get to the gym, you can take inspiration from a man who lived 3,000 years ago.

OVERCOMING THE POWER OF NOW: THE ULYSSES CONTRACT

This man was in a more extreme version of the gym scenario. He had something he wanted to do, but knew he wouldn't be able to resist temptation when the time came. For him it wasn't about getting a better physique; it was about saving his life from a group of mesmerizing maidens.

This was the legendary hero Ulysses, on his way back from triumph in the Trojan War. At some point on his long journey home, he realized that his ship would soon be passing an island where the beautiful Sirens lived. The Sirens were famous for singing songs so melodious that sailors were rapt and enchanted. The problem was that the sailors found the women irresistible, and would crash their ships into the rocks trying to get to them.

Ulysses desperately wanted to hear the legendary songs, but he didn't want to kill himself and his crew. So he hatched a plan. He knew that when he heard the music, he would be unable to resist steering toward the island's rocks. The problem wasn't the present rational Ulysses, but instead the future, illogical Ulysses—the person he'd become when the Sirens came within earshot. So Ulysses ordered his men to lash him securely to the mast of the ship. They filled their ears with beeswax so as not to hear the Sirens, and they rowed under strict orders to ignore any of his pleas and cries and writhing.

Ulysses knew that his future self would be in no position to make good decisions. So the Ulysses of sound mind arranged things so that he couldn't do the wrong thing. This sort of deal between your present and future self is known as a Ulysses contract.

In the case of going to the gym, my simple Ulysses

contract is to arrange in advance for a friend to meet me there: the pressure to uphold the social contract lashes me to the mast. When you start looking for them, you'll see that Ulysses contracts are all around you. Take college students who swap Facebook passwords during the week of their final exams; each student changes the password of the other so that neither can log on until finals are over. The first step for alcoholics in rehabilitation programs is to clear all the alcohol from their home, so the temptation is not in front of them when they're feeling weak. People with weight problems sometimes get surgery to reduce their stomach volume so they physically cannot overeat. In a different twist on a Ulysses contract, some people arrange things so that a violation of their promise will trigger a financial donation to an "anti-charity." For example, a woman who fought for equal rights her whole life wrote out a large check to the Ku Klux Klan, with strict orders to her friend to mail the check if she smoked another cigarette.

In all these cases, people structure things in the present so that their future selves can't misbehave. By lashing ourselves to the mast we can get around the seduction of the now. It's the trick that lets us behave in better alignment with the kind of person we would like to be. The key to the Ulysses contract is recognizing that we are different people in different contexts. To make better decisions, it's important not only to know yourself but all of your selves.

THE INVISIBLE MECHANISMS OF DECISION MAKING

Knowing yourself is only part of the battle—you also have to know that the outcome of your battles will not be the same every time. Even in the absence of a Ulysses contract, sometimes you'll feel more enthusiastic about going to the gym, and sometimes less so. Sometimes you're more capable of good decision making, and other times your neural parliament will come out with a vote you later regret. Why? It's because the outcome depends on many changing factors about the state of your body, states which can change hour to hour. For example: two men serving a prison sentence are scheduled to appear before a parole board. One prisoner comes before the board at 11:27 am. His crime is fraud and he's serving thirty months. Another prisoner appears at 1:15 pm. He has committed the same crime, for which he had been given the same sentence.

The first prisoner is denied parole; the second is granted parole. Why? What influenced the decision? Race? Looks? Age?

A study in 2011 analyzed a thousand rulings from judges, and found it likely wasn't about any of those factors. It was mostly about hunger. Just after the parole board had enjoyed a food break, a prisoner's chance of parole rose to its highest point of 65 percent. But a prisoner seen toward the end of a session had the lowest chances: just a 20 percent likelihood of a favorable outcome.

In other words, decisions get reprioritized as other needs rise in importance. Valuations change as circumstances change. A prisoner's fate is irrevocably intertwined with the judge's neural networks, which operate according to biological needs.

Some psychologists describe this effect as "ego-depletion," meaning that higher-level cognitive areas involved in executive function and planning (for example, the prefrontal cortex) get fatigued. Willpower is a limited resource; we run low on it, just like a tank of fuel. In the case of the judges, the more cases they had to make decisions about (up to thirty-five in one sitting) the more energy-depleted their brains became. But after eating something like a sandwich and a piece of fruit, their energy stores were refueled and different drives had more power in steering decisions.

Traditionally, we assume that humans are rational decision makers: they absorb information, process it, and come up with an optimal answer or solution. But real humans don't operate this way. Even judges, striving for freedom from bias, are imprisoned in their biology.

Our decisions are equally influenced when it comes to how we act with our romantic partners. Consider the choice of monogamy—bonding and staying with a single partner. This would seem like a decision that involves your culture, values, and morals. All that is true, but there's a deeper force acting on your decision making as well: your hormones. One in particular, called oxytocin, is a key ingredient in the magic of

WILLPOWER, A FINITE RESOURCE

We spend plenty of energy cajoling ourselves into making decisions we feel we ought to. To stay on the straight and narrow, we often fall back on willpower: that inner strength which allows you to pass on the cookie (or at least the second cookie), or which allows you to hit a deadline when you really want to be out in the sunshine. We all know what it feels like when our willpower feels run-down: after a long, hard day at work, people often find themselves making poorer choices—for example, eating a larger meal than they intended to, or watching television instead of hitting their next deadline.

So psychologist Roy Baumeister and colleagues put it to a closer test. People were invited to watch a sad movie. Half were told to react as they normally would, while the other half were instructed to suppress their emotions. After the movie, they were all given a hand exerciser and asked to squeeze it for as long as they could. Those who had suppressed their emotions gave up sooner. Why? Because self-control requires energy, which means we have less energy available for the next thing we need to do. And that's why resisting temptation, making hard decisions, or taking initiative all seem to draw from the same well of energy. So willpower isn't something that we just exercise—it's something we deplete.

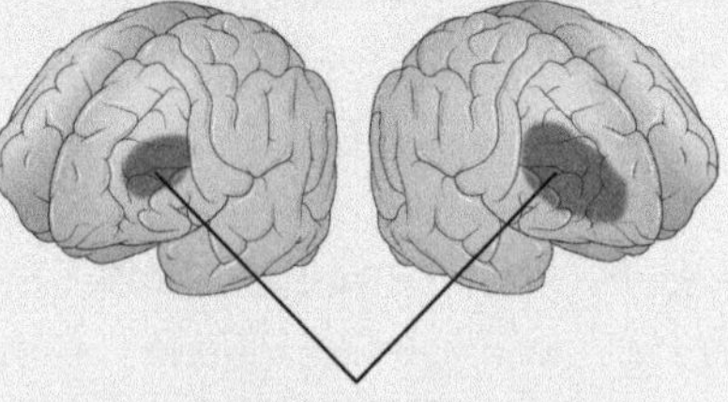

The dorsolateral prefrontal cortex becomes active when dieters choose the healthier food options in front of them, or when people choose to forego a small reward now for a better outcome later.

bonding. In one recent study, men who were in love with their female partners were given a small dose of extra oxytocin. They were then asked to rate the attractiveness of different women. With the extra oxytocin, the men found their partners more attractive—but not other women. In fact, the men kept a bit more physical distance from an attractive female research associate in the study. Oxytocin increased bonding to their partner.

Why do we have chemicals like oxytocin steering us toward bonding? After all, from an evolutionary perspective, we might expect that a male shouldn't want monogamy if his biological mandate is to spread his genes as widely as possible. But for the survival of the children, having two parents around is better than one. This simple fact is so important that the brain possesses hidden ways to influence your decision making on this front.

DECISIONS AND SOCIETY

A better understanding of decision making opens the door to better social policy. For example, each of us, in our own way, struggles with impulse control. At the extreme, we can end up as slaves to the immediate cravings of our impulses. From this perspective, we can gain a more nuanced understanding of social endeavors such as the War on Drugs.

Drug addiction is an old problem for society, leading to crime, diminished productivity, mental illness, disease transmission—and, more recently, to a burgeoning prison population. Nearly seven out of ten prisoners meet the criteria for substance abuse or dependence. In one study, 35.6 percent of convicted inmates were under the influence of drugs at the time of their criminal offense. Drug abuse translates into many tens of billions of dollars, mostly in terms of drug-related crime.

Most countries deal with the problem of drug addiction by criminalizing it. A few decades ago, 38,000 Americans were in prison for drug-related offenses. Today, it's half a million. On the face of it, that might sound like success in the War on Drugs—but this mass incarceration hasn't slowed the drug trade. This is because, for the most part, the people behind bars aren't the cartel bosses, or the mafia dons, or the big-time dealers—instead, the prisoners have been locked up for possession of a small amount of drugs, usually less than two grams. They're the users. The addicts. Going to prison doesn't solve their problem—it generally worsens it.

The US has more people in prison for drug-related crimes than the European Union has prisoners. The problem is that incarceration triggers an expensive and vicious cycle of relapse and reimprisonment. It breaks people's existing social circles and employment opportunities, and gives them new social circles and new employment opportunities—ones that typically fuel their addiction.

Every year the US spends $20 billion on the War on Drugs; globally, the total is over $100 billion. But the investment hasn't worked. Since the war began, drug use has expanded. Why hasn't the expenditure succeeded? The difficulty with drug supply is that it's like a water balloon: if you push it down in one place, it comes up somewhere else. Instead of attacking supply, the better strategy is to address demand. And drug demand is in the brain of the addict.

Some people argue that drug addiction is about poverty and peer pressure. Those do play a role, but at the core of the issue is the biology of the brain. In laboratory experiments, rats will self-administer drugs, continually hitting the delivery lever at the expense of food and drink. The rats aren't doing that because of finances or social coercion. They're doing it because the drugs tap into fundamental reward circuitry in their brains. The drugs effectively tell the brain that this decision is better than all the other things it could be doing. Other brain networks may join the battle, representing all the reasons to resist the drug. But in an addict, the craving network wins. The majority of drug addicts want to quit but find themselves unable. They end up becoming slaves to their impulses.

Because the problem with drug addiction lies in the brain, it's plausible that the solutions lie there too. One approach is to tip the balance of impulse control. This can be achieved by ramping up the certainty and swiftness of punishment—for instance, by requiring drug offenders to undergo twice-weekly drug testing, with automatic, immediate jail time for failure—thereby not relying on distant abstraction alone. Similarly, some economists propose that the drop in American crime since the early 1990s has been due, in part, to the increased presence of police on the streets. In the language of the brain, the police visibility stimulates the networks that weigh long-term consequences.

In my laboratory, we're working on another potentially effective approach. We are giving real-time feedback during brain imaging, allowing cocaine addicts to view their own brain activity and learn how to regulate it.

Meet one of our participants, Karen. She is bubbly and intelligent, and at fifty years old she retains a youthful energy. She's been addicted to crack cocaine for over two decades, and she describes the drug as having ruined her life. If she sees the drug right in front of her, she feels no choice but to take it. In ongoing experiments in my lab, we put Karen into the brain scanner (functional magnetic resonance imaging, or fMRI). We show her pictures of crack cocaine, and ask her to crave. That's easy for her to do, and it activates particular regions of her brain that we summarize as the craving network. Then we ask her to suppress her craving. We ask her to think about the cost crack cocaine has had to her—in terms of finances, in terms of relationships, in terms of employment. That activates a different set of brain areas, which we summarize as the suppression network. The craving and suppression networks are always battling it out for supremacy, and whichever wins at any moment determines what Karen does when offered crack.

Using fast computational techniques in the scanner, we can measure which network is winning: the short-term thinking of the craving network, or the long-term thinking of the impulse control or suppression network. We give Karen real-time visual feedback in the form of a speedometer so she can see how that battle is going. When her craving is winning, the needle is in the red zone; as she successfully suppresses, the needle moves to the blue zone. She can then use different approaches to discover what works to tip the balance of these networks.

By practicing over and over, Karen gets better at understanding what she needs to do to move the needle. She may or may not be consciously aware of how she's doing it, but by repeated practice she can strengthen the neural circuitry that allows her to suppress. This technique is still in its infancy, but the hope is that when she's next offered crack she'll have the cognitive tools to overcome her immediate cravings if she wants to. This training does not force Karen to behave in any particular way; it simply gives her the cognitive skills to have more control over her choice, rather than to be a slave to her impulses.

Drug addiction is a problem for millions of people. But prisons aren't the place to solve the problem. Equipped with an understanding of how human brains actually make decisions, we can develop new approaches beyond punishment. As we come to better appreciate the operations inside our brains, we can better align our behavior with our best intentions.

More generally, a familiarity with decision making can improve aspects of our criminal justice system well beyond addiction, ushering in policies which are more humane and cost-effective. What might that look like? It would begin with an emphasis on rehabilitation over mass incarceration. This may sound illusory, but in fact there are places already pioneering such an approach with great success. One such place is Mendota Juvenile Treatment Center in Madison, Wisconsin.

Many of the twelve- to seventeen-year-olds at Mendota have committed crimes that might otherwise qualify them for life in prison. Here, it qualifies them for admission. For many of the children, this is their last chance. The program started in the early 1990s to provide a new approach to working with youths the system had given up on. The program pays particular attention to their young, developing brains. As we saw in Chapter 1, without a fully developed prefrontal cortex, decisions are often made impulsively, without meaningful consideration of future consequences. At Mendota, this viewpoint illuminates an approach to rehabilitation. To help the children improve their self-control, the program provides a system of mentoring, counseling, and rewards. An important technique is to train them to pause and consider the future outcome of any choice they might make—encouraging them to run simulations of what might happen—thereby strengthening neural connections that can override the immediate gratification of impulses.

Poor impulse control is a hallmark characteristic of the majority of criminals in the prison system. Many people on the wrong side of the law generally know the difference between right and wrong actions, and they understand the threat of the punishment—but they are hamstrung by poor impulse control. They see an older woman with an expensive purse, and they don't pause to consider other options besides taking advantage of the opportunity. The temptation in the now overrides any consideration of the future.

While our current style of punishment rests on a bedrock of personal volition and blame, Mendota is an experiment in alternatives. Although societies possess deeply ingrained impulses for punishment, a different kind of criminal justice system—one with a closer relationship to the neuroscience of decisions—can be imagined. Such a legal system wouldn't let anyone off the hook, but it would be more concerned with how to deal with law breakers with an eye toward their future rather than writing them off because of their past. Those who break the social contracts need to be off the streets for the safety of society—but what happens in prison does not have to be based only on bloodlust, but also on evidence-based, meaningful rehabilitation.

Decision making lies at the heart of everything: who we are, what we do, how we perceive the world around us. Without the ability to weigh alternatives, we would be hostages to our most basic drives. We wouldn't be able to wisely navigate the now, or plan our future lives. Although you have a single identity, you're not of a single mind: instead, you are a collection of many competing drives. By understanding how choices battle it out in the brain, we can learn to make better decisions for ourselves, and for our society.

5

DO I NEED YOU?

What does your brain need to function normally? Beyond the nutrients from the food you eat, beyond the oxygen you breathe, beyond the water you drink, there's something else, something equally as important: it needs other people. Normal brain function depends on the social web around us. Our neurons require other people's neurons to thrive and survive.

HALF OF US IS OTHER PEOPLE

Over seven billion human brains traffic around the planet today. Although we typically feel independent, each of our brains operates in a rich web of interaction with one another—so much so that we can plausibly look at the accomplishments of our species as the deeds of a single, shifting megaorganism.

Brains have traditionally been studied in isolation, but that approach overlooks the fact that an enormous amount of brain circuitry has to do with other brains. We are deeply social creatures. From our families, friends, coworkers, and business partners, our societies are built on layers of complex social interactions. All around us we see relationships forming and breaking, familial bonds, obsessive social networking, and the compulsive building of alliances.

All of this social glue is generated by specific circuitry in the brain: sprawling networks that monitor other people, communicate with them, feel their pain, judge their intentions, and read their emotions. Our social skills are deeply rooted in our neural circuitry—and understanding this circuitry is the basis of a young field of study called social neuroscience.

Take a moment to consider how different the following items are: bunnies, trains, monsters, airplanes, and children's toys. As different as they are, these can all be the main characters in popular animated films, and we have no difficulty in assigning intentions to them. A viewer's brain needs very few hints to take on the assumption that these characters are like us, and therefore we can laugh and cry over their escapades.

This penchant to assign intention to nonhuman characters was highlighted in a short film made in 1944 by psychologists Fritz Heider and Marianne Simmel. Two simple shapes—a triangle and circle—come together and spin around one another. After a moment, a larger triangle comes lurking into the scene. It bumps up against and pushes the smaller triangle. The circle slowly sneaks back into a rectangular structure and closes it behind; meanwhile, the large triangle chases the smaller triangle away. The large triangle

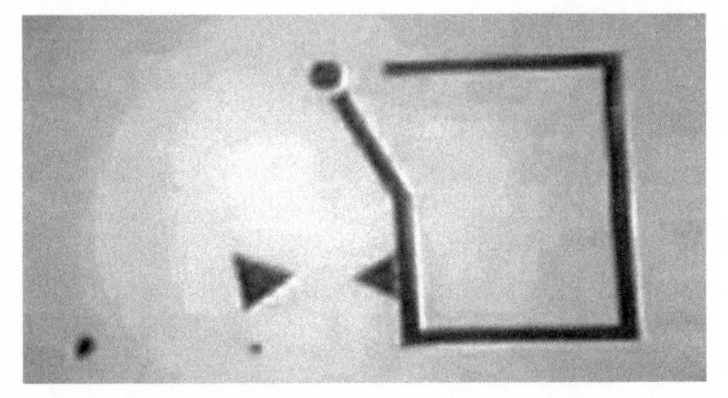

People irresistibly impose a narrative on moving shapes.

then comes to the door of the structure, menacingly. The triangle pries the door open and comes in after the circle, who frenetically (and unsuccessfully) looks for other ways to escape. Just when the situation looks its darkest, the little triangle returns. He pulls open the door and the circle dashes out to meet him. Together they shut the door behind them, trapping the large triangle inside. Penned in, the large triangle smashes against the walls of the structure. Outside, the little triangle and circle spin around one another.

When people watched this short film and were asked to describe what they saw, you might expect that they described simple shapes moving around. After all, it's just a circle and two triangles changing coordinates.

But that's not what the viewers reported. They described a love story, a fight, a chase, a victory. Heider and Simmel used this animation to demonstrate how readily we perceive social intention all around us. Moving shapes hit our eyes, but we see meaning and motives and emotion, all in the form of a social narrative. We can't help but impose stories. From time immemorial, people have watched the flights of birds, the movement of stars, the swaying of trees, and invented stories about them, interpreting them as having intention.

This kind of storytelling is not just a quirk; it's an important clue into brain circuitry. It unmasks the degree to which our brains are primed for social interaction. After all, our survival depends on quick assessments of who is friend and who is foe. We navigate the social world by judging other people's intentions. Is she trying to be helpful? Do I need to worry about him? Are they looking out for my best interests?

Our brains make social judgments constantly. But do we learn this skill from life experience, or are we born with it? To find out, one can investigate whether babies have it. Reproducing an experiment from psychologists Kiley Hamlin, Karen Wynn, and Paul Bloom at Yale University, I invited babies, one at a time, to a puppet show.

These babies are less than a year old, just beginning to explore the world around them. They're all short on life experience. They're positioned on their mothers' laps to watch the show. When the curtain parts, a duck struggles to open a box with toys in it. The duck grasps at the lid but just can't get a good grip on it. Two bears, wearing two different-colored shirts, watch.

After a few moments, one of the bears helps the duck, working with him to grip the side of the box and pry the lid open. They hug momentarily, and then the lid closes again.

Now the duck tries to get the lid open again. The other bear, watching, throws his weight onto the lid, preventing the duck from succeeding.

That's the whole show. In a short, wordless plot, one bear has been helpful to the duck, and the other bear has been mean.

When the curtain falls, and then reopens, I take both bears and carry them over to the watching baby. I hold them up, indicating to the child to choose one of them to play with. Remarkably, as was found by the Yale researchers, almost all the babies choose the bear that was kind. These babies can't walk or talk, but they already have the tools to make judgments about others.

It's often assumed that trustworthiness is something we learn to assess, based on years of experience in the world. But simple experiments like these demonstrate that, even as babies, we come equipped with social antennae for feeling our way through the world. The brain comes with inborn instincts to detect who's trustworthy, and who isn't.

THE SUBTLE SIGNALS AROUND US

As we grow, our social challenges become more subtle and complex. Beyond words and actions, we have to interpret inflection, facial expressions, body language. While we are consciously concentrating on what we are discussing, our brain machinery is busy processing complex information. The operations are so instinctive that they're essentially invisible.

Often, the best way to appreciate something is to see what the world looks like when it's missing. For a man named John Robison, the normal activity of the social brain was something he was simply unaware of as he grew up. He was bullied and rejected by other children but found a love of machines. As he describes it, he could spend time with a tractor and it wouldn't tease him. "I guess I learned how to make friends with

AUTISM

Autism is a neurodevelopmental disorder which affects 1 percent of the population. Although it's established that both genetic and environmental causes underpin its development, the number of individuals diagnosed with autism has been on the rise in recent years, with little to no evidence explaining this increase. In people not affected by autism, many regions of the brain are involved in searching for social cues about the feelings and thoughts of others. In autism, this brain activity is not seen as strongly—and this is paralleled by diminished social skills.

the machines before I made friends with other people," he says.

In time, John's affinity for technology took him to places his bullies could only dream of. By twenty-one, he was a roadie for the band KISS. However, even while surrounded by legendary rock and roll excess, his outlook remained different from others'. When people would ask him about the different musicians and what they were like, John would respond by explaining how they had played Sun Coliseum with seven base amps chained together. He would explain that there were 2,200 watts in the bass system, and could enumerate the amplifiers and what the crossover frequencies were. But he couldn't tell you a thing about the musicians who sang through them. He lived in a world of technology and equipment. It wasn't until he was forty that John was diagnosed with Asperger's, a form of autism.

Then something happened that transformed John's life. In 2008 he was invited to take part in an experiment at Harvard Medical School. A team led by Dr. Alvaro Pascual-Leone was using transcranial magnetic stimulation (TMS) to assess how activity in one area of the brain affected activity in another area. TMS emits a strong magnetic pulse next to the head, which in turn induces a small electric current in the brain, temporarily disrupting local brain activity. The experiment was meant to help the researchers gain greater knowledge about the autistic brain. The team used TMS to target different regions of John's brain involved in higher-order cognitive function. At first, John reported the stimulation had no effect. But in one session, the researchers applied TMS to the dorsolateral prefrontal cortex, an evolutionarily recent part of the brain involved in flexible thinking and abstraction. John reported that he somehow became different.

John called up Dr. Pascual-Leone to let him know that the effects of the stimulation seemed to have "unlocked" something in him. The effects lasted beyond the experiment itself, John reported. For John it had opened up a whole new window on to the social world. He simply didn't realize that there were messages emanating from the facial expressions of other people—but after the experiment, he was now aware of those messages. To John, his experience of the world was now changed. Pascual-Leone was skeptical. He figured if the effects were real they wouldn't last, given that the effects of TMS typically persist only a few minutes to hours. Now, although Pascual-Leone does not fully understand what happened, he allows that the stimulation seems to have fundamentally changed John.

In the social realm, John went from experiencing black and white to full color. He now sees a communication channel that he was never able to detect before. John's story isn't simply about hope for new treatment techniques for autism spectrum disorder. It reveals the importance of the unconscious machinery running under the hood, every moment of our waking lives, devoted to social connection—brain circuitry that continuously decodes the emotions of others based on subtle facial, auditory, and other sensory cues.

"I knew that people could display signs of crazed anger," he says. "But if you asked about more subtle expressions—like, I think you're sweet or I wonder what you're hiding or I'd really like to do that or I wish you'd do this—I had no idea about things like that."

Every moment of our lives, our brain circuitry decodes the emotions of others based on extremely subtle facial cues. To better understand how we read faces so rapidly and automatically, I invited a group of people to my lab. We placed two electrodes on their faces—one on the forehead and one on the cheek—to measure small changes in their expressions. Then we had them look at photographs of faces.

When participants looked at a photo that showed, say, a smile, or a frown, we were able to measure short periods of electrical activity that indicated their own facial muscles were moving, often very subtly. This is because of something called mirroring: they were automatically using their own facial muscles to copy the expressions they were seeing. A smile was reflected by a smile, even if the movement of their muscles was too slight to be visually obvious. Without meaning to, people ape one another.

This mirroring sheds light on a strange fact: couples who are married for a long time begin to resemble each other, and the longer they've been married, the stronger the effect. Research suggests this is not simply because they adopt the same clothes or hairstyles, but because they've been mirroring each other's faces for so many years that their patterns of wrinkles start to look the same.

Why do we mirror? Does it serve a purpose? To find out, I invited a second group of people to the lab—similar to the first group, except for one thing: this new group of people had been exposed to the most lethal toxin on the planet. If you were to ingest even a few drops of this neurotoxin, your brain could no longer command your muscles to contract, and you would die from paralysis (specifically, your diaphragm would no longer be able to move, and you would suffocate). Given these facts, it seems unlikely that people would pay to have this injected into themselves. But they do. This is the Botulinum toxin, derived from a bacterium, and it's commonly marketed under the brand name Botox. When injected into facial muscles, it paralyzes them and thereby reduces wrinkling.

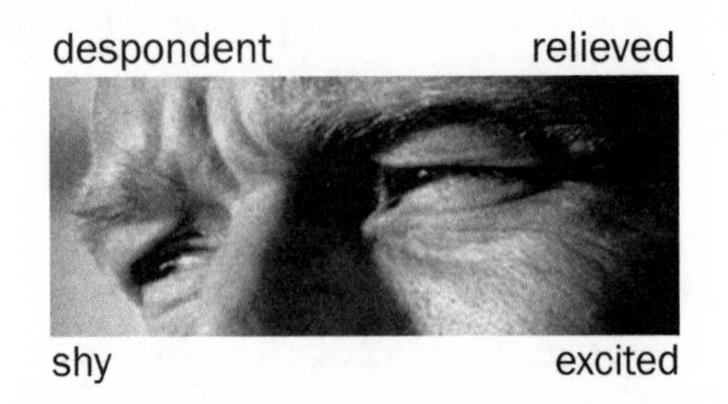

In the Reading the Mind in the Eyes test (Baron-Cohen et al., 2001), participants are shown thirty-six photographs of facial expressions, each accompanied by four words.

However, beyond the cosmetic benefit, there's a less known side effect of Botox. We showed Botox users the same set of photos. Their facial muscles showed less mirroring on our electromyogram. No surprise there—their muscles have been purposely weakened. The surprise was something else, originally reported in 2011 by David Neal and Tanya Chartrand. Similar to their original experiment, I asked participants from both groups (Botox and non-Botox) to look at expressive faces and to choose which of four words best described the emotion shown.

On average, those with Botox were worse at identifying the emotions in the pictures correctly. Why? One hypothesis suggests that the lack of feedback from their facial muscles impaired their ability to read other people. We all know that the less mobile faces of Botox users can make it hard to tell what they're feeling; the surprise is that those same frozen muscles can make it hard for them to read others.

Here's a way to think about this result: my facial muscles reflect what I'm feeling, and your neural machinery takes advantage of that. When you're trying to understand what I'm feeling, you try on my facial expression. You don't mean to do it—it happens rapidly and unconsciously—but that automatic mirroring of my expression gives you a rapid estimate of what I'm likely to be feeling. This is a powerful trick for your brain to gain a better understanding of me and make better predictions about what I'll do. As it turns out, it's just one trick of many.

THE JOYS AND SORROWS OF EMPATHY

We go to the movies to escape into worlds of love and heartbreak and adventure and fear. But the heroes and villains are just actors projected in two dimensions on a screen—so why should we care at all about what happens to those fleeting phantasms? Why do movies make us weep, laugh, gasp?

To understand why you care about the actors, let's begin with what happens in your brain when you are in pain. Imagine that someone stabs your hand with a syringe needle. There's no single place in the brain where that pain is processed. Instead, the event activates several different areas of the brain, all operating in concert. This network is summarized as the pain matrix.

Here's the surprising part: the pain matrix is crucial to how we connect with others. If you watch somebody else get stabbed, most of your pain matrix becomes activated. Not those areas that tell you you've actually been touched, but instead those parts involved in the emotional experience of pain. In other words, watching someone else in pain and being in pain use the same neural machinery. This is the basis of empathy.

To empathize with another person is to literally feel their pain. You run a compelling simulation of what it would be like if you were in that situation. Our capacity for this is why stories—like movies and novels—are so absorbing and so pervasive across human culture. Whether it's about total strangers or made-up characters, you experience their agony and their ecstasy. You fluidly become them, live their lives, and stand in their vantage points. When you see another person suffer, you can try to tell yourself that it's their issue, not yours—but neurons deep in your brain can't tell the difference.

This built-in facility to feel another person's pain is part of what makes us so good at stepping out of our shoes and into their shoes, neurally speaking. But why do we have this facility in the first place? From an evolutionary point of view, empathy is a useful skill: by gaining a better grasp of what someone is feeling, it gives a better prediction about what they'll do next.

However, the accuracy of empathy is limited, and

in many cases we simply project ourselves onto others. Take as an example Susan Smith, a mother in South Carolina who in 1994 kindled the empathy of a nation when she reported to the police that she had been carjacked by a man who drove away with her sons still in the car. For nine days, she pled on national television for the rescue and return of her boys. Strangers around the nation offered help and support. Eventually, Susan Smith confessed to the murder of her own children. Everyone had fallen for her story of the carjacking, because her real act was so outside the realm of normal predictions. Although the details of her case are all reasonably obvious in retrospect, they were difficult to see at the time—because we typically interpret other people from the vantage point of who we are and what we're capable of.

We can't help but simulate others, connect with others, care about others, because we're hardwired to be social creatures. That raises a question. Are our brains dependent on social interaction? What would happen if the brain were starved of human contact?

In 2009, peace activist Sarah Shourd and her two companions were hiking in the mountains of Northern Iraq—an area that was, at that time, peaceful. They followed recommendations from locals to see the Ahmed Awa waterfall. Unfortunately, this waterfall was located at the Iraqi border with Iran. They were arrested by Iranian border guards on suspicion of being American spies. The two men were put in the same cell, but Sarah was separated from them in solitary confinement. With the exception of two thirty-minute periods each day, she spent the next 410 days in an isolated cell.

In Sarah's words:

> *In the early weeks and months of solitary confinement you're reduced to an animal-like state. I mean, you are an animal in a cage, and the majority of your hours are spent pacing. And the animal-like state eventually transforms into a more plant-like state: your mind starts to slow down and your thoughts become repetitive. Your brain turns on itself and becomes the source of your worst pain and your worst torture. I'd relive every moment of my life, and eventually you run out of memories. You've told them all to yourself so many times. And it doesn't take that long.*

Sarah's social deprivation caused deep psychological pain: without interaction, a brain suffers. Solitary confinement is illegal in many jurisdictions, precisely because observers have long recognized the damage caused by stripping away one of the most vital aspects of a human life: interaction with others. Starved of contact with the world, Sarah rapidly entered a hallucinatory state:

> *The sun would come in at a certain time of day at an angle through my window. And all of the little dust particles in my cell were illuminated by*

the sun. I saw all those particles of dust as other human beings occupying the planet. And they were in the stream of life, they were interacting, they were bouncing off one another. They were doing something collective. I saw myself as off in a corner, walled up. Out of the stream of life.

In September 2010, after more than a year in captivity, Sarah was released and allowed to rejoin the world. The trauma of the event stayed with her: she suffered from depression and was easily led to panic. The next year she married Shane Bauer, one of the other hikers. She reports that she and Shane are able to calm one another, but it's not always easy: they both carry emotional scars.

The philosopher Martin Heidegger suggested that it is difficult to speak of a person "being," instead we are typically "being in the world." This was his way of emphasizing that the world around you is a large part of who you are. The self doesn't exist in a vacuum.

Although scientists and clinicians can observe what happens to people in solitary confinement, it is difficult to study directly. However, an experiment by neuroscientist Naomi Eisenberger can give insight into what happens in the brain in a slightly tamer condition: when we are excluded from a group.

Imagine throwing a ball around with a couple of other people, and at some point you get cut out of the game: the other two throw back and forth between themselves, excluding you. Eisenberger's experiment is based on that simple scenario. She had volunteers play a simple computer game in which their animated character threw a ball around with two other players. The volunteers were led to believe that the other players were controlled by two other humans, but in fact they were just part of a computer program. At first, the others played nicely—but after a while, they cut the volunteer out of the game, and simply threw between each other.

Eisenberger had the volunteers play this game while they were lying down in a brain scanner (the technique is called functional magnetic resonance imaging, or fMRI—see Chapter 4). She found something remarkable: when the volunteers were left out of the game, areas involved in their pain matrix became active. Not getting the ball might seem insignificant, but to the brain social rejection is so meaningful that it hurts, literally.

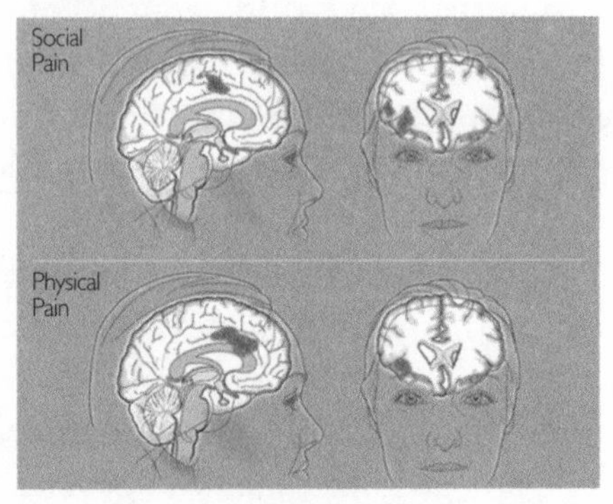

Social pain—such as that resulting from exclusion—activates the same brain regions as physical pain.

Why does rejection hurt? Presumably, this is a clue that social bonding has evolutionary importance—in other words, the pain is a mechanism that steers us toward interaction and acceptance by others. Our inbuilt neural machinery drives us toward bonding with others. It urges us to form groups.

This sheds light on the social world that surrounds us: everywhere, humans constantly form groups. We bind together through links of family, friendship, work, style, sports teams, religion, culture, skin pigment, language, hobbies, and political affiliation. It gives us comfort to belong to a group—and that fact gives us a critical hint about our species' history.

BEYOND SURVIVAL OF THE FITTEST

When we think about human evolution, we're all familiar with the concept of survival of the fittest: it calls to mind the picture of a strong and wily individual who can outfight, outrun, or outmate other members of its species. In other words, one has to be a good competitor to thrive and survive. That model has good explanatory power, but it leaves some aspects of our behavior difficult to explain. Consider altruism: why does survival of the fittest explain why people help each other out? Selection of the strongest individual doesn't seem to cover it, so theorists introduced the additional idea of "kin selection." This means that I care not only about myself, but also others with whom

I share genetic material, for example, brothers and cousins. As the evolutionary biologist J. S. Haldane quipped, "I would gladly jump in a river to save two of my brothers, or eight of my cousins."

However, even kin selection is not enough to explain all the facets of human behavior, because people get together and cooperate irrespective of kinship. That observation leads to the idea of "group selection." Here's the concept: if a group is composed entirely of people who cooperate, everyone in the group will be better off for it. On average, you'll fare better than other people who aren't very cooperative with their neighbors. Together, the members of a group can help each other to survive. They're safer, more productive, and better able to overcome challenges. This drive to bond with others is called eusociality (eu is Greek for good), and it provides a glue, irrespective of kinship, that allows the building of tribes, groups, and nations. It's not that individual selection doesn't occur; it's just that it doesn't provide the complete picture. Although humans are competitive and individualistic much of the time, it's also the case that we spend quite a bit of our lives cooperating for the good of the group. This has allowed human populations to thrive across the planet, and to build societies and civilizations—feats that individuals, no matter how fit, could never pull off in isolation. Real progress is only possible with alliances that become confederations, and our eusociality is one of the major factors in the richness and complexity of our modern world.

So our drive to come together into groups yields a survival advantage—but it has a dark side, as well. For every ingroup, there must exist at least one outgroup.

OUTGROUPS

An understanding of ingroups and outgroups is critical to understand our history. Repeatedly, all across the globe, groups of people inflict violence on other groups, even those that are defenseless and pose no direct threat. The year 1915 saw the systematic killing of over a million Armenians by the Ottoman Turks. In the Nanking massacre of 1937, the Japanese invaded China and killed hundreds of thousands of unarmed civilians. In 1994, in a period of one hundred days, the Hutus in Rwanda killed 800,000 Tutsis, largely with machetes.

I don't view this with the detached eye of a historian. If you were to look at my family tree, you would see that most of the branches come to an abrupt end in the early 1940s. They were murdered because they were Jewish, caught in the jaws of the Nazi genocide as a scapegoated outgroup.

After the Holocaust, Europe got in the habit of vowing "never again." But fifty years later, genocide happened again—this time just 600 miles away, in Yugoslavia. Between 1992 and 1995, during the Yugoslav War, over 100,000 Muslims were slaughtered by Serbians in violent acts that became known as "ethnic cleansing." One of the worst events of the war happened in Srebrenica: here, over the course of ten days, 8,000 Bosnian Muslims—known as Bosniaks—were shot and killed. They had taken refuge inside a United Nations compound after Srebrenica was surrounded by siege forces, but on July 11, 1995, the United Nations commanders expelled all the refugees from the compound, delivering them into the hands of their enemies waiting just outside the gates. Women were raped, men were executed, and even children were killed.

I flew to Sarajevo to better understand what had happened, and there I had the chance to speak with a tall, middle-aged man named Hasan Nuhanović. Hasan, a Bosnian Muslim, had been working at the compound as a UN translator. His family was also there, among the refugees, but they had been sent out of the compound to die, while only he had been allowed to stay because of his value as a translator. His mother, father, and brother were killed that day. The part that haunts him the most is this: "the continuation of the killings, of torture, was perpetrated by our neighbors—the very people we had been living with for decades. They were capable of killing their own school friends."

To exemplify the ways in which normal social interaction broke down, he told me how Serbs arrested a Bosniak dentist. They hung him by his arms from a lightpole, and they beat him with a metal bar until they broke his spine. Hasan told me how the dentist hung there for three days while Serbian children walked

SYNDROME E

What allows a diminished emotional reaction to harming another person? The neurosurgeon Itzhak Fried points out that when you look across violent events all over the world, you find the same character of behavior everywhere. It's as though people shift from their normal brain function to act in a specific way. In the same way a physician can look for coughing and fever with pneumonia, he suggested that one can look for and identify particular behaviors that characterize perpetrators in violent situations—and he named this "Syndrome E." In Fried's framework, Syndrome E is characterized by a diminished emotional reactivity, which allows repetitive acts of violence. It also includes hyperarousal, or as the Germans call it, *Rausch*—a feeling of elation in doing these acts. There's group contagion: everybody's doing it, and it catches on and spreads. There's compartmentalization, in which somebody can care about his own family, and yet perform violence on someone else's family.

From a neuroscientific point of view, the important clue is that other brain functions, such as language and memory and problem solving, are intact. That suggests it's not a brain-wide change, but instead only involves areas involved in emotion and in empathy. It's as though they become, in effect, short-circuited: they no longer participate in decision making. Instead, a perpetrator's choices are now fueled by parts of the brain that underpin logic and memory and reasoning and so on, but not the networks that involve emotional consideration of what it is like to be someone else. In Fried's view, this equates to moral disengagement. People are no longer using the emotional systems that under normal circumstances steer their social decision making.

past his body on their way to school. As he put it: "There are universal values and these values are very basic: don't kill. In April 1992, this 'don't kill' suddenly disappeared—and it became 'go and kill.' "

What allows such an alarming shift in human interaction? How can it be compatible with a eusocial species? Why does genocide continue to happen all around our planet? Traditionally we examine warfare and killings in the context of history and economics and politics. However, for a complete picture, I believe we need also to understand this as a neural phenomenon. It would normally feel unconscionable to murder your neighbor. So what suddenly allows hundreds or thousands of people to do exactly that? What is it about certain situations that short-circuits the normal social functioning of the brain?

SOME MORE EQUAL THAN OTHERS

Can a breakdown of normal social functioning be studied in the laboratory? I designed an experiment to find out.

Our first question was a simple one: does your basic sense of empathy toward someone change depending on whether they are in your ingroup or outgroup?

We put participants in the scanner. They saw six hands on the screen. Like a spinning wheel in a game show, the computer randomly picks one of the hands. That hand then expands into the middle of the screen, and you watch it get touched with a cotton swab, or stabbed with a syringe needle. These are two actions that yield about the same activity in the visual system, but very different reactions in the rest of the brain.

As we saw earlier, watching someone else in pain activates one's own pain matrix. That's the basis of empathy. So now we were able to push our questions about empathy to the next level. Once we had established this baseline condition, we made a very simple change: the same six hands appeared on the screen, but now each had a one-word label, reading Christian, Jewish, Atheist, Muslim, Hindu, or Scientologist. When a hand was randomly selected, it expanded to the middle of the screen and was then touched with the cotton swab or stabbed with the syringe needle. Our experimental question was this: would your brain care as much when seeing a member of an outgroup getting hurt?

We found a good deal of individual variability, but on average, people's brains showed a larger empathic response when they saw someone in their ingroup in pain, and less of a response when it was a member of one of their outgroups. The result is especially remarkable given that these were simply one-word labels: it takes very little to establish group membership.

A basic categorization is enough to change your brain's preconscious response to another person in pain. Now, one might have opinions about the divisiveness of religion, but there's a deeper point to note here: in our study, even atheists showed a larger response to pain in the hand labeled "atheist," and less of an empathic response to other labels. So the result is not fundamentally about religion—it's about which team you're on.

We see that people can feel lower empathy for members of an outgroup. But to understand something like violence or genocide, we still need to drill down one step further, to dehumanization.

Lasana Harris of the University of Leiden in Holland has conducted a series of experiments that move us closer to understanding how that happens. Harris is looking for changes in the brain's social network, in particular the medial prefrontal cortex (mPFC). This region becomes active when we're interacting with, or thinking about, other people—but it's not active when we're dealing with inanimate objects, like a coffee mug.

Harris shows volunteers photographs of people from different social groups, for example, homeless people, or drug addicts. And he finds that the mPFC is less active when they look at a homeless person. It's as though the person is more like an object.

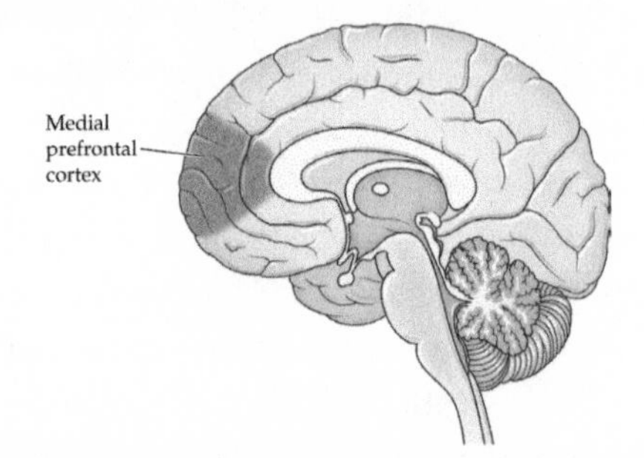

The medial prefrontal cortex is involved in thinking about other people—at least, most other people.

As he puts it, by shutting down the systems that see the homeless person as a fellow human, one doesn't have to experience the unpleasant pressures of feeling bad about not giving money. In other words, the homeless have become dehumanized: the brain is viewing them more like objects and less like people. Not surprisingly, one is less likely to treat them with consideration. As Harris explains: "if you don't properly diagnose people as human beings, then the moral rules that are reserved for human people may not apply."

Dehumanization is a key component of genocide. Just as the Nazis viewed the Jews as something less than human, the Serbs in former Yugoslavia viewed the Muslims this way.

When I was in Sarajevo, I walked along the main street. During the war it became known as Snipers' Alley because civilian men, women, and children were killed by riflemen crouched in the surrounding hillsides and neighboring buildings. This street became one of the most powerful symbols of the horror of the war. How does a normal city street come to that?

This war, like all others, was fueled by an effective form of neural manipulation, one that's been practiced for centuries: propaganda. During the Yugoslav war the main news network, Radio Television of Serbia, was controlled by the Serb government and consistently presented distorted news stories as factual. The network made up reports of ethnically motivated attacks by Bosnian Muslims and Croats against the Serb people. They continually demonized Bosnians and Croatians, and used negative language in their descriptions of Muslims. At the height of bizarreness, the network broadcast an unfounded story that Muslims were feeding Serbian children to the hungry lions of the Sarajevo zoo.

Genocide is only possible when dehumanization happens on a massive scale, and the perfect tool for this job is propaganda: it keys right into the neural networks that understand other people, and dials down the degree to which we empathize with them.

We've seen that our brains can be manipulated by political agendas to dehumanize other people, which can then lead to the darkest side of human acts. But is it possible to program our brains to prevent this? One possible solution lies in a 1960s experiment that was conducted not in a science lab, but in a school.

It was 1968, the day after the assassination of civil rights leader Martin Luther King. Jane Elliott, a teacher in a small town in Iowa, decided to demonstrate to her class what prejudice was about. Jane asked her class whether they knew how it would feel to be judged by the color of their skin. The students mostly thought they could. But she wasn't so sure, so she launched what was destined to become a famous experiment. She announced that the blue-eyed people were "the better people in this room."

Jane Elliott: The brown-eyed people do not get to use the drinking fountain. You'll have to use the paper cups. You brown-eyed people are not to play with the blue-eyed people on the playground, because you are not as good as blue-eyed people. The brown-eyed people in this room today are going to wear collars. So that we can tell from a distance what color your eyes are. On page 127 . . . Is everyone ready? Everyone but Laurie. Ready, Laurie?

Child: She's a brown-eye.

Jane: She's a brown-eye. You'll begin to notice today that we spend a great deal of time waiting for brown-eyed people.

A moment later, Jane looks around for her yardstick, and two boys pipe up. Rex points out to her where the yardstick is, and Raymond helpfully offers, "Hey, Mrs. Elliott, you better keep that on your desk so if the brown people [sic], the brown-eyed people get out of hand."

I recently sat down with those two boys, now grown men: Rex Kozak and Ray Hansen. They both have blue eyes. I asked them if they remembered what their behavior was like on that day. Ray reported that "I was tremendously evil to my friends. I was going out of my way to pick on my brown-eyed friends, for the sake of my own promotion." He recalled that at that time his hair was quite blond and his eyes were quite blue, "and I was the perfect little Nazi. I looked for ways to be mean to my friends, who minutes or hours earlier had been very close to me."

The next day, Jane reversed the experiment. She announced to the class:

The brown-eyed people may take off their collars. And each of you may put your collar on a blue-eyed person. The brown-eyed people get five extra minutes of recess. You blue-eyed people are not allowed to be on the playground equipment at any time. You blue-eyed people are not to play with the brown-eyed people. Brown-eyed people are better than blue-eyed people.

Rex described what the reversal was like: "It takes your world and shatters it like you've never had your world shattered before." When Ray was in the down group, he felt such a deep sense of loss, of personality, and of self, that he felt it was almost impossible to function.

One of the most important things we learn as humans is perspective taking. And children don't typically get a meaningful exercise in that. When one is forced to understand what it's like to stand in someone else's shoes, it opens up new cognitive pathways. After the exercise in Mrs. Elliott's classroom, Rex was more vigilant against racist statements; he remembers telling his father, "that's not appropriate." Rex remembers that moment fondly: he felt affirmed by it, and he knew he'd begun to change as a person.

The brilliance of the blue eyes/brown eyes exercise was that Jane Elliott switched which group was on top. That allowed the children to extract a larger lesson: systems of rules can be arbitrary. The children learned that the truths of the world aren't fixed, and moreover they're not necessarily truths. This exercise empowered the children to see through the smoke and mirrors of political agendas, and to form their own opinions—surely a skill we would want for all our children.

Education plays a key role in preventing genocide. Only by understanding the neural drive to form ingroups and outgroups—and the standard tricks by which propaganda plugs into this drive—can we hope to interrupt the paths of dehumanization that end in mass atrocity.

In this age of digital hyperlinking, it's more important than ever to understand the links between humans. Human brains are fundamentally wired to interact: we're a splendidly social species. Although our social drives can sometimes be manipulated, they also sit squarely at the center of the human success story.

You might assume that you end at the border of your skin, but there's a sense in which there's no way to mark the end of you and the beginning of all those around you. Your neurons and those of everyone on the planet interplay in a giant, shifting superorganism. What we demarcate as you is simply a network in a larger network. If we want a bright future for our species, we'll want to continue to research how human brains interact—the dangers as well as the opportunities. Because there's no avoiding the truth etched into the wiring of our brains: we need each other.

WHO WILL WE BE?

The human body is a masterpiece of complexity and beauty—a symphony of forty trillion cells working in concert. However, it has its limitations. Your senses set boundaries on what you can experience. Your body sets limits on what you can do. But what if the brain could understand new kinds of inputs and control new kinds of limbs—expanding the reality we inhabit? We're at a moment in human history when the marriage of our biology and our technology will transcend the brain's limitations. We can hack our own hardware to steer a course into the future. This is poised to fundamentally change what it will mean to be a human.

Over the last 100,000 years our species has been on quite a journey: we've gone from living as primitive hunter-gatherers surviving on scraps to a planet-conquering hyperconnected species that defines its own destiny. Today we enjoy mundane experiences that our ancestors could never have dreamed of. We have clean rivers that we can call into our well-adorned caves when we desire. We hold small rock-sized devices that contain the knowledge of the world. We regularly see the tops of clouds and the curvature of our home planet from space. We send messages to the other side of the globe in eighty milliseconds and upload files to a floating space colony of humans at sixty megabits per second. Even when simply driving to work, we routinely move at speeds that outstrip biology's great masterpieces, such as cheetahs. Our species owes its runaway success to the special properties of the three pounds of matter stored inside our skulls.

What is it about the human brain that has made this journey possible? If we can understand the secrets behind our achievements, then perhaps we can direct the brain's strengths in careful, purposeful ways, opening a new chapter in the human story. What do the next thousand years have in store for us? In the far future, what will the human race be like?

A FLEXIBLE, COMPUTATIONAL DEVICE

The secret to understanding our success—and our future opportunity—is the brain's tremendous ability to adjust, known as brain plasticity. As we saw in Chapter 2, this feature has allowed us to drop into any environment and pick up on the local details we need to survive, including the local language, local environmental pressures, or local cultural requirements.

Brain plasticity is also the key to our future, because it opens the door to making modifications to our own hardware. Let's begin by understanding just how flexible a computational device the brain is. Consider the case of a young girl named Cameron Mott. At the age of four she began to have violent seizures. The seizures were aggressive: Cameron would suddenly drop to the floor, requiring her to wear a helmet all the time. She was quickly diagnosed with a rare and debilitating disease called Rasmussen's Encephalitis. Her neurologists knew that this form of epilepsy would lead to paralysis and eventually to death—and so they proposed a drastic surgery. In 2007, in an operation that took almost twelve hours, a team of neurosurgeons removed an entire half of Cameron's brain.

What would be the long-term effects of removing half her brain? As it turns out, the consequences were surprisingly slight. Cameron is weak on one side of her body, but otherwise she's essentially indistinguishable from the other children in her class. She has no problems understanding language, music, math, stories. She's good in school and she participates in sports.

How could this be possible? It's not that one half of Cameron's brain was simply not needed; instead, the remaining half of Cameron's brain dynamically rewired to take over the missing functions, essentially cramming all the operations into half the brain space. Cameron's recovery underscores a remarkable ability of the brain: it rewires itself to adjust to the inputs, outputs, and tasks at hand.

In this critical way, the brain is fundamentally unlike the hardware in our digital computers. Instead, it's "liveware." It reconfigures its own circuitry. Although the adult brain isn't quite as flexible as a child's, it still retains an astonishing ability to adapt and change. As we saw in previous chapters, every time we learn something new, whether it's the map of London or the ability to stack cups, the brain changes itself. It's this property of the brain—its plasticity—that enables a new marriage between our technology and our biology.

PLUGGING IN PERIPHERAL DEVICES

We've become progressively better at plugging machinery directly into our bodies. You may not realize it, but currently hundreds of thousands of people are walking around with artificial hearing and artificial vision.

With a device called a cochlear implant, an external microphone digitizes a sound signal and feeds it to the auditory nerve. Similarly, the retinal implant digitizes a signal from a camera, and sends it through an electrode grid plugged into the optic nerve at the back of the eye. For deaf and blind people around the planet, these devices have restored their senses.

It wasn't always clear that such an approach would work. When these technologies were first introduced, many researchers were skeptical: the brain is wired up with such precision and specificity that it wasn't clear there could be a meaningful dialogue between metal electrodes and biological cells. Would the brain be able to understand crude, nonbiological signals, or would it be confused by them?

As it turns out, the brain learns to interpret the signals. Getting used to these implants is a bit like learning a new language for the brain. At first the foreign electrical signals are unintelligible, but the neural networks eventually extract patterns in incoming data. Although the input signals are crude, the brain finds a way to make sense of them. It hunts for patterns, cross-referencing with other senses. If there's structure to be found in the incoming data, the brain ferrets it out—and after several weeks the information begins to take on meaning. Even though the implants give slightly different signals than do our natural sense organs, the brain figures out how to make do with the information it can get.

ARTIFICIAL HEARING AND VISION

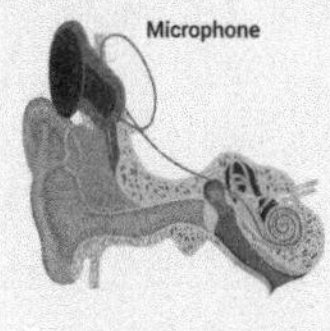

Cochlear implant

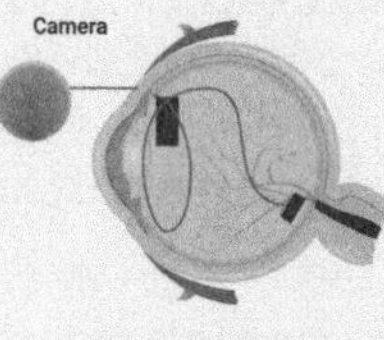

Retinal implant

A cochlear implant bypasses problems in the biology of the ear and feeds its audio signals directly to the undamaged auditory nerve, the brain's data cable for sending electrical impulses on to the auditory cortex for decoding. The implant picks up sounds from the outside world and passes them to the auditory nerve by means of sixteen tiny electrodes. The experience of hearing doesn't arrive immediately: people have to learn to interpret the foreign dialect of the signals fed to the brain. As one cochlear implant recipient, Michael Chorost, describes his experience:

"When the device was turned on a month after surgery, the first sentence I heard sounded like 'Zzzzzz szz szvizzz ur brfzzzzzz?' My brain gradually learned how to interpret the alien signal. Before long, 'Zzzzzz szz szvizzz ur brfzzzzzz?' became 'What did you have for breakfast?' After months of practice, I could use the telephone again, even converse in loud bars and cafeterias."

Retinal implants work on similar principles. The tiny electrodes of the retinal implant bypass the normal functions of the photoreceptor sheet, sending out their tiny sparks of electrical activity. These implants are used mostly for eye diseases in which the photoreceptors at the back of the eye are degenerating, but in which the cells of the optic nerve remain healthy. Even though the signals sent by the implant are not precisely what the visual system is used to, the downstream processes are able to learn to extract the information they need for vision.

PLUG AND PLAY: AN EXTRASENSORY FUTURE

The brain's plasticity allows new inputs to be interpreted. What sensory opportunities does that open up?

We come into the world with a standard set of basic senses: hearing, touch, sight, smell, and taste, along with other senses such as balance, vibration, and temperature. The sensors we have are the portals by which we pick up signals from our environment.

However, as we saw in the first chapter, these senses only allow us to experience a tiny fraction of the world around us. All the information sources for which we don't have sensors are invisible to us.

I think of our sensory portals as peripheral plug-and-play devices. The key is that the brain doesn't know and doesn't care where it gets the data. Whatever information comes in, the brain figures out what to do with it. In this framework, I think of the brain as a general-purpose computing device: it operates on whatever it's fed. The idea is that Mother Nature only needed to invent the principles of brain operation once—and then she was freed up to tinker with designing new input channels.

The end result is that all these sensors we know and love are merely devices that can be swapped in and out. Stick them in and the brain can get to work. In this framework, evolution doesn't need to continually redesign the brain, just the peripherals, and the brain figures out how to utilize them.

Just look across the animal kingdom, and you'll find a mind-boggling variety of peripheral sensors in use by animal brains. Snakes have heat sensors. The glass knifefish has electrosensors for interpreting changes in the local electrical field. Cows and birds have magnetite, with which they can orient themselves to the Earth's magnetic field. Animals can see in ultraviolet; elephants can hear at very long distances, while dogs experience a richly scented reality. The crucible of natural selection is the ultimate hacker space, and these are just some of the ways that genes have figured out how to channel data from the outside world into the internal world. The end result is that evolution has built a brain that can experience many different slices of reality.

The consequence I want to highlight is that there may be nothing special or fundamental about the sensors we're used to. They're just what we've inherited from a complex history of evolutionary constraints. We're not stuck with them.

Our main proof of principle for this idea comes from a concept called sensory substitution, which refers to feeding sensory information through unusual sensory channels such as vision through touch. The brain figures out what to do with the information, because it doesn't care how the data finds its way in.

Sensory substitution might sound like science fiction, but in fact it's already well established. The first demonstration was published in the journal *Nature* in 1969. In that report, neuroscientist Paul

Bach-y-Rita demonstrated that blind subjects could learn to "see" objects—even when the visual information was fed to them in an unusual way. Blind people were seated in a modified dental chair, and the video feed from a camera was converted into a pattern of small plungers pressed against their lower back. In other words, if you put a circle in front of the camera, the participant would feel a circle on her back. Put a face in front of the camera, and the participant feels the face on her back. Amazingly, blind people could come to interpret the objects, and could also experience the increasing size of approaching objects. They could, at least in a sense, come to see through their backs.

This was the first example of sensory substitution of many to follow. Modern incarnations of this approach include turning a video feed into a sound stream, or a series of small shocks on the forehead or the tongue.

An example of the latter is the postage stamp–sized device called the BrainPort, which works by delivering tiny electrical shocks to the tongue via a small grid that sits on the tongue. A blind subject wears sunglasses with a small camera attached. Camera pixels are converted into electrical pulses on the tongue, which feels something like the fizz of a carbonated drink. Blind people can become quite good at using the BrainPort, navigating obstacle courses or throwing a ball into a basket. One blind athlete, Erik Weihenmayer, uses the BrainPort to rock climb, assessing the position of crags and crevices from the patterns on his tongue.

If it sounds crazy to "see" through your tongue, just keep in mind that seeing is never anything but electrical signals streaming into the darkness of your skull. Normally this happens via the optic nerves, but there's no reason the information can't stream in via other nerves instead. As sensory substitution demonstrates, the brain takes whatever data comes in and figures out what it can make of it.

One of the projects in my laboratory is to build a platform for enabling sensory substitution. Specifically, we have built a wearable technology called the Variable Extra-Sensory Transducer, or VEST. Worn inconspicuously under the clothing, the VEST is covered with tiny vibratory motors. These motors convert data streams into dynamic patterns of vibration across the torso. We're using the VEST to give hearing to the deaf.

After about five days of using the VEST, a person who was born deaf can correctly identify spoken words. Although the experiments are still in their early stages, we expect that after several months of wearing the VEST, users will come to have a direct perceptual experience—essentially the equivalent of hearing.

It may seem strange that a person can come to hear via moving patterns of vibration on the torso. But just as with the dental chair or the tongue grid, the trick is this: the brain doesn't care how it gets the information, as long as it gets it.

SENSORY AUGMENTATION

Sensory substitution is great for circumventing broken sensory systems—but beyond substitution, what if we could use this technology to extend our sensory inventory? To this end, my students and I are currently adding new senses to the human repertoire to augment our experience of the world.

Consider this: the Internet is streaming petabytes of interesting data, but currently we can only access that information by staring at a phone or computer screen. What if you could have real-time data streamed into your body, so that it became part of your direct experience of the world? In other words, what if you could feel data? This could be weather data, stock exchange data, Twitter data, cockpit data from an airplane, or data about the state of a factory—all encoded as a new vibratory language that the brain learns to understand. As you went about your daily tasks, you could have a direct perception of whether it's raining a hundred miles away or whether it's going to snow tomorrow. Or you could develop intuitions about where the stock markets were going, subconsciously identifying the movements of the global economy. Or you could sense what's trending across the Twittersphere, and in this way be tapped into the consciousness of the species.

Although this sounds like science fiction, we're not far off from this future—all thanks to the brain's talent at extracting patterns, even when we're not trying.

THE VEST

To provide sensory substitution for the deaf, my graduate student Scott Novich and I built the VEST. This wearable tech captures sound from the environment and maps it to small vibrational motors all over the torso. The motors activate in patterns according to the frequencies of the sound. In this way, sound becomes moving patterns of vibrations.

At first, these vibratory signals make no sense. But with enough practice, the brain works out what to do with the data. Deaf people become able to translate the complicated patterns on the torso into an understanding of what's being said. The brain figures out how to unconsciously unlock the patterns, similar to the manner in which a blind person comes to effortlessly read Braille.

The VEST has the potential to be a game changer for the deaf community. Unlike a cochlear implant, it doesn't require an invasive surgery. And it's at least twenty times cheaper, which makes it a solution that can be global.

The bigger vision for the VEST is this: beyond sound, it can also serve as a platform for any kind of streaming information to find its way to the brain.

See videos of the VEST in action at eagleman.com.

That is the trick that can allow us to absorb complex data and incorporate it into our sensory experience of the world. Like reading this page, absorbing new data streams will come to feel effortless. Unlike reading, however, sensory addition would be a way to take on new information about the world without having to consciously attend to it.

At the moment, we don't know the limits—or if there are limits—to the kinds of data the brain can incorporate. But it's clear that we are no longer a natural species that has to wait for sensory adaptations on an evolutionary timescale. As we move into the future, we will increasingly design our own sensory portals on the world. We will wire ourselves into an expanded sensory reality.

HOW TO GET A BETTER BODY

How we sense the world is only half the story. The other half is how we interact with it. In the same way that we are beginning to modify our sensory selves, can the brain's flexibility be leveraged to modify the way we reach out and touch the world?

Meet Jan Scheuermann. Because of a rare genetic disease called spinocerebellar disorder, the spinal cord nerves connecting her brain to her muscles have deteriorated. She can feel her body, but she can't move it. As she describes it, "my brain is saying 'lift up' to my arm, but the arm is saying 'I can't hear you.'" Her total paralysis made her an ideal candidate for a new study at the University of Pittsburgh School of Medicine.

Researchers there have implanted two electrodes into her left motor cortex, the last stop for brain signals before they plunge down the spinal cord to control the muscles of the arm. The electrical storms in her cortex are monitored, translated on a computer to understand the intention, and the output is used to control the world's most advanced robotic arm.

When Jan wants to move the robotic arm, she simply thinks about moving it. As she moves the arm, Jan tends to talk to it in the third person: "Go up. Go down, down, down. Go right. And grasp. Release." And the arm does so on cue. Although she speaks the commands out loud, she has no need to. There's a direct physical link between her brain and the arm. Jan reports that her brain has not forgotten how to move an arm, even though it hadn't moved one in ten years. "It's like riding a bicycle," she says.

Jan's proficiency points to a future in which we use technology to enhance and extend our bodies, not only replacing limbs or organs, but improving them: elevating them from human fragility to something more durable. Her robotic arm is just the first hint of an upcoming bionic era in which we'll be able to control much stronger and longer-lasting equipment than the skin and muscle and brittle bones we're born with. Among other things, that opens up new possibilities for space travel, something for which our delicate bodies are ill-equipped.

Beyond replacement limbs, advancing brain-machine interface technology suggests more exotic possibilities. Imagine extending your body to be something unrecognizable. Start with this idea: what if you could use your brain signals to wirelessly control a machine across the room? Envision answering e-mails while simultaneously using your motor cortex to control a thought-controlled vacuum cleaner. At first glance, the concept may sound unworkable, but keep in mind that brains are great at running tasks in the background, not requiring much in the way of conscious bandwidth. Just consider how easily you can drive a car while simultaneously talking to a passenger and fiddling with the radio knob.

With the proper brain-machine interface and wireless technology, there's no reason you couldn't control large devices such as a crane or a forklift wirelessly, at a distance, with your thoughts, in the same way that you might absentmindedly dig with a trowel or play a guitar. Your capacity to do this well would be enhanced by sensory feedback, which could be done visually (you watch how the machine moves), or even by feeding data back into your somatosensory cortex (you feel how the machine moves). Controlling such limbs would take practice and be awkward at first, in the same way that a baby has to flail for some months to learn how to finely control its arms and legs. With time, these machines would effectively become an extra limb—one that could have extraordinary strength, hydraulic or otherwise. They would come to feel the way that your arms or legs do to you now. They would just be another limb—simple extensions of ourselves.

We don't know of a theoretical limit on the kinds of signals the brain could learn to incorporate. It may be possible to have almost any sort of physical body and any kind of interaction with the world that we want. There's no reason an extension of you couldn't be taking care of tasks on the other side of the planet, or mining rocks on the moon while you're enjoying a sandwich here on Earth.

The body we arrive with is really just the starting point for humanity. In the distant future, we won't just be extending our physical bodies, but fundamentally our sense of self. As we take on new sensory experiences and control new kinds of bodies, that will change us profoundly as individuals: our physicality sets the stage for how we feel, how we think, and who we are. Without the limitations of the standard-issue senses and the standard-issue body, we'll become different people. Our great-great-great-great-grandchildren may struggle to understand who we were, and what was important to us. At this moment in history, we may have more in common with our Stone Age ancestors than with our near-future descendants.

STAYIN' ALIVE

We're already beginning to extend the human body, but no matter how much we enhance ourselves, there is one snag that's difficult to avoid: our brains and bodies are built of physical stuff. They will deteriorate and die. There will come a moment when all your neural activity will come to a halt, and then the glorious experience of being conscious will come to an end. It doesn't matter who you know or what you do: this is the fate of all of us. In fact, it's the fate of all life, but only humans are so unusually foresighted that we suffer over this knowledge.

Not everyone is content to suffer; some have chosen to fight death's specter. Scattered confederacies of researchers are interested in the idea that a better understanding of our biology can address our mortality. What if in the future we didn't have to die?

When my friend and mentor, Francis Crick, was cremated, I spent some time thinking about what a shame it was that all his neural matter was going up in flames. That brain contained all the knowledge, wisdom, and intellect of one of the heavyweight champions of twentieth-century biology. All the archives of his life—his memories, his capacity for insight, his sense of humor—were stored in the physical structure of his brain, and simply because his heart had stopped everyone was content to throw away the hard drive. It made me wonder: could the information in his brain be preserved somehow? If the brain were preserved, could a person's thoughts and awareness and personhood ever be brought back to life?

For the past fifty years, the Alcor Life Extension Foundation has been developing technology they believe will allow people living today to enjoy a second life cycle later. The organization currently stores 129 people in a deep freeze that halts their biological decay.

Here's how cryopreservation works: first, an interested party signs his life insurance policy over to the foundation. Then, upon the legal declaration of his death, Alcor is alerted. A local team sweeps in to manage the body.

The team immediately transfers the body to an ice bath. In a process known as cryoprotective perfusion, they circulate sixteen different chemicals to protect the cells as the body cools. The body is then relocated as quickly as possible to the Alcor operating room for the final stage of the procedure. The body is cooled by computer-controlled fans circulating extremely low-temperature nitrogen gas. The goal is to cool all parts of the body below -124°C as rapidly as possible to avoid any ice formation. The process takes about three hours, at the end of which the body will have "vitrified," that is, reached a stable ice-free state. The body is then further cooled to -196°C over the next two weeks.

Not all clients choose to have their whole body frozen. A less expensive option is to simply preserve the head. The separation of the head from the body is performed on a surgical table, where the blood and fluids are washed out and, as with the whole-body clients, are replaced with liquids that fix the tissue into place.

At the end of the procedure, the clients are lowered into ultracooled liquid in giant stainless steel cylinders called dewars. This is where they'll remain for a long

LEGAL VERSUS BIOLOGICAL DEATH

A person is declared legally dead when either his brain is clinically dead or his body has experienced irreversible cessation of respiration and circulation. For the brain to be declared dead, all activity must have ceased in the cortex, involved in higher function. After brain death, vital functions can still be maintained for organ donation or body donation, a fact critical for Alcor. Biological death, on the other hand, happens in the absence of intervention, and involves the death of cells throughout the body: in the organs and in the brain, and means that the organs are no longer suitable for donation. Without oxygen from circulating blood, the body's cells rapidly start to die. To preserve a body and a brain in its least degraded form, cell death must be stopped, or at least decelerated, as quickly as possible. In addition, during cooling the priority is to prevent ice crystals from forming, which can destroy the delicate structures of the cells.

time; no one on the planet today knows how to successfully unfreeze and reanimate these frozen residents. But that's not the point. The hope is that one day the technology will exist to carefully thaw—and then revive—the people in this community. Civilizations in the distant future, it is presumed, will command the technology to cure the diseases that ravaged these bodies and brought them to a halt.

Alcor members understand that the technology may never exist to revive them. Each person dwelling in the Alcor dewars took a leap of faith, hoping and dreaming that someday the technology will materialize to thaw them out, revive them, and give them a second chance at life. The venture is a gamble that the future will develop the necessary technology. I spoke to a member of the community (who awaits his eventual entry into the dewars when the time comes), and he allowed the whole conception was a wager. But, he pointed out, at least it gives him a better-than-zero chance of cheating death—better odds than the rest of us.

Dr. Max More, who runs the facility, doesn't use the word "immortality." Instead, he says, Alcor is about giving people a second chance at life, with the potential to live thousands of years or longer. Until that time comes, Alcor is their final resting place.

DIGITAL IMMORTALITY

Not everyone with a penchant for life extension has a fondness for cryopreservation. Others have moved along a different line of inquiry: what if there were other ways to access the information stored in a brain? Not by bringing a deceased person back to life, but instead by finding a way to read out the data directly. After all, the submicroscopically detailed structure of your brain contains all your knowledge and memories—so why couldn't that book be decrypted?

Let's take a look at what would be required to do that. To begin, we'd need extraordinarily powerful computers to store the detailed data of an individual brain. Fortunately, our exponentially growing computational power hints at profound possibilities. Over the past twenty years, computing power has increased over a thousand times. The processing power of computer chips has doubled approximately every eighteen months, and this trend continues. The technologies of our modern era allow us to store unimaginable amounts of data and run gargantuan simulations.

Given our computing potential, it seems likely that we'll someday be able to scan a working copy of the human brain onto a computer substrate. There is nothing, in theory, that precludes this possibility. However, the challenge needs to be realistically appreciated.

The typical brain has about eighty-six billion neurons, each making about ten thousand connections. They connect in a very specific manner, unique to each person. Your experiences, your memories, all the stuff that makes you you is represented by the unique pattern of the quadrillion connections between your brain cells. This pattern, far too large to comprehend, is summarized as your "connectome." In an ambitious endeavor, Dr. Sebastian Seung at Princeton is working with his team to excavate the fine details of a connectome.

With a system this microscopic and complex, it's inordinately difficult to map out the network of connectivity. Seung uses serial electron microscopy, which involves making a series of very thin slices of brain tissue using an extremely precise blade. (At the moment, mouse brains are used, not human.) Each slice is subdivided into tiny areas, and each of these is scanned by an extraordinarily powerful electron microscope. The result of each scan is a picture known as an electron micrograph—and this represents a segment of brain magnified one hundred thousand times. At this resolution it's possible to make out fine features of the brain.

Once these slices are stored in the computer, the more difficult work begins. One very thin slice at a time, the borders of the cells are traced out—traditionally by hand, but increasingly by computer algorithms. Then the images are stacked atop one another, and an attempt is made to connect the full extent of individual cells across slices, to reveal them in their three-dimensional richness. In this painstaking manner a model emerges, revealing what is connected to what.

The dense spaghetti of connections is just a few billionths of a meter across, about the size of the head

THE PACE OF TECHNOLOGICAL CHANGE

In 1965, Gordon Moore, cofounder of the computing giant Intel, made a prediction about the rate of progress in computing power. "Moore's Law" forecast that as transistors became smaller and more precise, the number that could fit onto a computer chip would double every two years, exponentially increasing computing power over time. Moore's prediction has held true through the intervening decades, and has become shorthand for the exponentially accelerating pace of technological change. Moore's Law is used by the computing industry to guide long-term planning and set goals for technological advancement. Because the law predicts that technological progress will increase exponentially rather than linearly, some predict that at today's rate there will be 20,000 years' worth of progress in the next hundred years. At this pace we can expect to see radical advancements in the technology that we rely on.

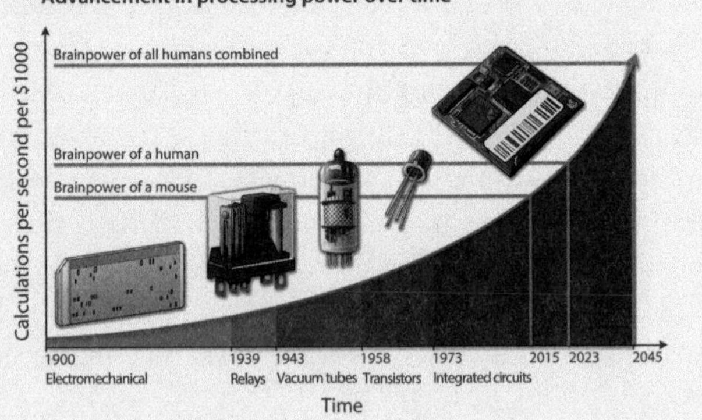

of a pin. It's not difficult to see why reconstructing the full picture of all the connections in a human brain is such a daunting task, and one that we have no real hope of accomplishing anytime soon. The amount of data required is gargantuan: to store a high-resolution architecture of a single human brain would require a zettabyte of capacity. That's the same size as all the digital content of the planet right now.

Throwing far into the future, let's imagine that we could get a scan of *your* connectome. Would that information be enough to represent you? Could this snapshot of all the circuitry of your brain actually have consciousness—*your* consciousness? Probably not. After all, the circuit diagram (which shows us what connects to what) is only half of the magic of a functioning brain. The other half is all the electrical and chemical activity that runs on top of those connections. The alchemy of thought, of feeling, of awareness—this emerges from quadrillions of interactions between brain cells every second: the release of chemicals, the changes in the shapes of proteins, the traveling waves of electrical activity down the axons of neurons.

Consider the enormity of the connectome, and then multiply that by the vast number of things happening every second at every one of those connections, and you'll get a sense of the magnitude of the problem. Unfortunately for us, systems of this magnitude cannot be comprehended by the human brain. Fortunately for us, our computational power is moving in the right direction to eventually open up a possibility: a simulation of the system. The next challenge is not just reading it out, but making it run.

Such a simulation is exactly what a team of researchers at the École polytechnique fédérale de Lausanne (EPFL) in Switzerland is working toward. Their goal is to deliver by 2023 a software and hardware infrastructure capable of running a whole human brain simulation. The Human Brain Project is an ambitious research mission that collects data from neuroscience laboratories across the globe—this includes data on individual cells (their contents and structure) to connectome data to information about large-scale activity patterns in groups of neurons. Slowly, one experiment at a time, each new finding on the planet provides a tiny piece of a titanic puzzle. The goal of the Human Brain Project is to achieve a simulation of a brain that uses detailed neurons, realistic in their structure and their behavior. Even with this ambitious goal and over a billion euros of funding from the European Union, the human brain is still totally out of reach. The current goal is to build a simulation of a rat brain.

We are only at the beginning of our endeavor to map and simulate a full human brain, but there's no theoretical reason why we can't get there. But here's a key question: would a working simulation of the brain be conscious? If the details were captured and simulated correctly, would we be looking at a sentient being? Would it think and be self-aware?

SERIAL ELECTRON MICROSCOPY AND THE CONNECTOME

Signals from the environment are translated into electrochemical signals carried by brain cells. It is the first step by which the brain taps into information from the world outside the body.

Tracing the dense tangle of billions of interconnected neurons requires specialized technology, as well as the world's sharpest blade. A technique called "serial block-face scanning electron microscopy" generates high-resolution 3-D models of complete neural pathways from tiny slices of brain tissue. It's the first technique to yield 3-D images of the brain at nanoscale resolution (one-billionth of a meter).

Like a deli-slicer, a high-precision diamond blade mounted inside a scanning microscope cuts layer after layer from a tiny block of brain, producing a filmstrip in which each frame is an ultrathin slice. Each sliver is scanned by an electron microscope. The scans are then digitally layered on top of one another to create a high-resolution 3-D model of the original block.

By tracing features from slice to slice, a model emerges of the tangle of neurons that crisscross and intertwine. Given that an average neuron can be between 4–100 billionths of a meter in length and have 10,000 different branches, it's a formidable task. The challenge of mapping a full human connectome is expected to take several decades.

DOES CONSCIOUSNESS REQUIRE THE PHYSICAL STUFF?

In the same way that computer software can run on different hardware, it may be that the software of the mind can run on other platforms as well. Consider the possibility this way: what if there is nothing special about biological neurons themselves, and instead it's only how they communicate that makes a person who they are? That prospect is known as the computational hypothesis of the brain. The idea is that the neurons and synapses and other biological matter aren't the critical ingredients: it's the computations they happen to be implementing. It may be that what the brain physically is doesn't matter, but instead what it does.

If that turns out to be true, then in theory you could run the brain on any substrate. As long as the computations chug along in the right way, then all your thoughts, emotions, and complexities should arise as a product of the complex communications within the new material. In theory, you might swap cells for circuitry, or oxygen for electricity: the medium doesn't matter, provided that all the pieces and parts are connecting and interacting in the right way. In this way, we may be able to "run" a fully functioning simulation of you without a biological brain. According to the computational hypothesis, such a simulation would actually be you.

The computational hypothesis of the brain is just that—a hypothesis—one that we don't yet know is

RAT BRAINS

The rat has had a terrible reputation for much of human history, but to modern neuroscience the rat (and the mouse) plays a crucial role in many areas of research. Rats have larger brains than mice, but both have important similarities to the human brain—in particular, the organization of the cerebral cortex, the outer layer that's so important for abstract thinking.

The outer layer of the human brain, the cortex, is folded over on itself to allow more of it to be packed into the skull. If you flattened the average adult cortex out it would cover 2,500 square centimeters (a small tablecloth). The rat brain, in contrast, is completely smooth. Despite these obvious differences in appearance and size, there are fundamental similarities between the two brains at the cellular level.

Under a microscope it is almost impossible to tell the differences between a rat neuron and a human neuron. Both brains wire up in much the same way and go through the same developmental stages. Rats can be trained to do cognitive tasks—from distinguishing between scents to finding their way through a maze—and this allows researchers to correlate the details of their neural activity to specific tasks.

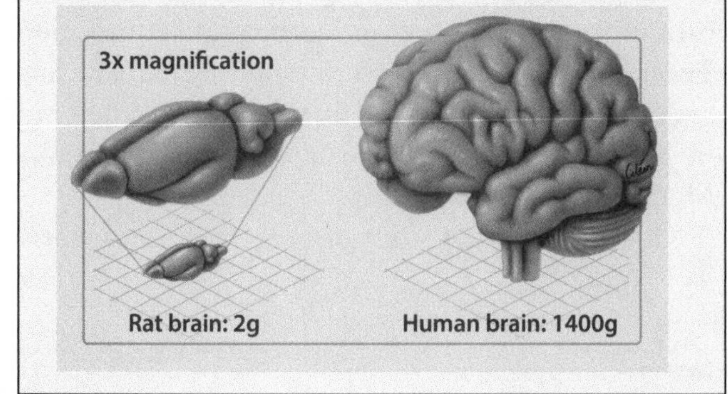

true. After all, there may be something special and undiscovered about the biological wetware, and in that case we're stuck with the biology we arrived with. However, if the computational hypothesis is correct, then a mind could live in a computer.

If it turns out to be possible to simulate a mind, that leads to a different question: do we have to copy the traditional biological way of doing it? Or might it be possible to create a different kind of intelligence, of our own invention, from scratch?

ARTIFICIAL INTELLIGENCE

People have been trying for a long time to create machines that think. That line of research—artificial intelligence—has been around since at least the 1950s. Although the initial pioneers were heady with optimism, the problem has turned out to be unexpectedly difficult. Although we'll soon have cars that drive themselves, and it's almost two decades since a computer first beat a chess grand master, the goal of a truly sentient machine still waits to be achieved. When I was a child, I expected that we would have robots interacting with us by now, taking care of us and engaging in meaningful conversation. The fact that we're still quite distant from that outcome speaks to the depth of the enigma of how the brain functions, and how far we still have to go to decode Mother Nature's secrets.

One of the latest attempts to create an artificial

intelligence can be found at the University of Plymouth, in England. It's called iCub, and it's a humanoid robot designed and engineered to learn like a human child. Traditionally, robots are preprogrammed with what they need to know about their tasks. But what if robots could develop the way a human infant does—by interacting with the world, by imitating and learning by example? After all, babies don't come into the world knowing how to speak and walk—but they come with curiosity and they pay attention and they imitate. Babies use the world they're in as a textbook to learn by example. Couldn't a robot do the same?

The iCub is about the size of a two-year-old. It has eyes and ears and touch sensors, and these allow it to interact with and learn about the world.

If you present a new object to iCub and name it ("this is a red ball"), the computer program correlates the visual image of the object with the verbal label. So the next time you present the red ball and ask "what is this?" it will answer "this is a red ball." The aim is that with each interaction, the robot continually adds to its base of knowledge. By making changes and connections within its internal code, it builds a repertoire of appropriate responses.

It often gets things wrong. If you present and name several objects and push iCub to name them all, you'll get several mistakes and a large number of "I don't know" responses. That's all part of the process. It also reveals how difficult it is to build intelligence.

I spent quite a bit of time interacting with iCub,

and it's an impressive project. But the longer I was there, the more it was obvious that there was no mind behind the program. Despite its big eyes and friendly voice and childlike movements, it becomes clear that iCub is not sentient. It's run by lines of code, not trains of thought. And even though we're still in the early days of AI, one can't help but chew on an old and deep question in philosophy: can lines of computer code ever come to think? While iCub can say "red ball," does it really experience redness or the concept of roundness? Do computers do just what they're programmed to do, or can they really have internal experience?

CAN A COMPUTER THINK?

Can a computer ever be programmed so that it has awareness, a mind? In the 1980s the philosopher John Searle came up with a thought experiment that gets right to the heart of this question. He called it the Chinese Room Argument.

It goes like this: I am locked in a room. Questions are passed to me through a small letter slot—and these messages are written only in Chinese. I don't speak Chinese. I have no clue at all what's written on these pieces of paper. However, inside this room I have a library of books, and they contain step-by-step instructions that tell me exactly what to do with these symbols. I look at the grouping of symbols, and I simply follow steps in the book that tell me what Chinese symbols to copy down in response. I write those on the slip of paper and pass it back out of the slot.

When the Chinese speaker receives my reply message, it makes sense to her. It seems as though whoever is in the room is answering her questions perfectly, and therefore it seems obvious that the person in the room must understand Chinese. I've fooled her, of course, because I'm only following a set of instructions, with no understanding of what's going on. With enough time and a big enough set of instructions I could answer almost any question posed to me in Chinese. But I, the operator, do not understand Chinese. I manipulate symbols all day long, but I have no idea what the symbols mean.

Searle argued this is just what is happening inside a computer. No matter how intelligent a program like iCub seems to be, it's only following sets of instructions to spit out answers—manipulating symbols without ever really understanding what it's doing.

Google is an example of this principle. When you send Google a query, it doesn't understand your question or its own answer: it simply moves around zeros and ones in logic gates and returns zeros and ones to you. With a mind-blowing program like Google Translate, I can speak a sentence of Swahili and it can return the translation in Hungarian. But it's all algorithmic. It's all symbol manipulation, just like the person inside the Chinese Room. Google Translate doesn't understand anything about the sentence; nothing carries any meaning to it.

The Chinese Room Argument suggests that as we develop computers that mimic human intelligence, they won't actually understand what they're talking about; there will be no meaning to anything they do. Searle used this thought experiment to argue that there's something about human brains that won't be explained if we simply analogize them to digital computers. There's a gap between symbols that have no meaning, and our conscious experience.

There's ongoing debate about the interpretation of the Chinese Room Argument, but however one construes it, the argument exposes the difficulty and the mystery of how physical pieces and parts ever come to equal our experience of being alive in the world. With every attempt to simulate or create a human-like intelligence, we're confronted by a central unsolved question of neuroscience: how does something as rich as the subjective feeling of being me—the sting of pain, the redness of red, the taste of grapefruit—arise from billions of simple brain cells running through their operations? After all, each brain cell is just a cell, following local rules, running its basic operations. By itself, it can't do much. So how do billions of these add up to the subjective experience of being me?

GREATER THAN THE SUM

In 1714, Gottfried Wilhelm Leibniz argued that matter alone could never produce a mind. Leibniz was a

German philosopher, mathematician, and scientist who is sometimes called "the last man who knew everything." To Leibniz, brain tissue alone could not have an interior life. He suggested a thought experiment, known today as Leibniz's Mill. Imagine a large mill. If you were to walk around inside of it, you would see its cogs and struts and levers all moving, but it would be preposterous to suggest that the mill is thinking or feeling or perceiving. How could a mill fall in love or enjoy a sunset? A mill is just made of pieces and parts. And so it is with the brain, Leibniz asserted. If you could expand the brain to the size of a mill and stroll around inside it, you would only see pieces and parts. Nothing would obviously correspond to perception. Everything would simply be acting on everything else. If you wrote down every interaction, it wouldn't be obvious where thinking and feeling and perceiving reside.

When we look inside the brain, we see neurons, synapses, chemical transmitters, electrical activity. We see billions of active, chattering cells. Where are you? Where are your thoughts? Your emotions? The feeling of happiness, the color of indigo blue? How can you be made of mere matter? To Leibniz, the mind seemed inexplicable by mechanical causes.

Is it possible that Leibniz overlooked something in his argument? By looking at the individual pieces and parts of a brain, he may have missed a trick. Maybe thinking about walking around in the mill is the wrong way to approach the question of consciousness.

CONSCIOUSNESS AS AN EMERGENT PROPERTY

To understand human consciousness, we may need to think not in terms of the pieces and parts of the brain, but instead in terms of how these components interact. If we want to see how simple parts can give rise to something bigger than themselves, look no farther than the nearest anthill.

With millions of members in a colony, leaf-cutter ants cultivate their own food. Just like humans, they're farmers. Some of the ants set forth from the nest to find fresh vegetation; when they find it, they chew off large pieces that they hump back to the nest. However, the ants don't eat these leaves. Instead, smaller worker ants take the pieces of leaves, chew them into smaller pieces, and use these as fertilizer to grow fungus in large underground "gardens." The ants feed the fungus, and the fungus blossoms into small fruiting bodies which the ants later eat. (The relationship has become so symbiotic that the fungus can no longer reproduce on its own; it relies entirely on the ants for its propagation.) Using this successful farming strategy, the ants build enormous nests underground, something spanning hundreds of square meters. Just like humans, they have perfected an agricultural civilization.

Here's the important part: although the colony is like a superorganism that accomplishes extraordinary feats, each ant individually behaves very simplistically. It just follows local rules. The queen doesn't give commanding orders; she doesn't coordinate the behavior from on high. Instead, each ant reacts to local chemical signals from other ants, larvae, intruders, food, waste, or leaves. Each ant is a modest, autonomous unit whose reactions depend only on its local environment and the genetically encoded rules for its variety of ant.

Despite the lack of centralized decision making, the leaf-cutter ant colonies exhibit what appears to be extraordinarily sophisticated behavior. (Beyond farming, they also accomplish feats like finding the maximum distance from all colony entrances to dispose of dead bodies, a sophisticated geometric problem.)

The important lesson is that the complex behavior of the colony doesn't arise from complexity in the individuals. Each ant doesn't know that it is part of a successful civilization: it just runs its small, simple programs.

When enough ants come together, a superorganism emerges—with collective properties that are more sophisticated than its basic parts. This phenomenon, known as "emergence," is what happens when simple units interact in the right ways and something larger arises.

What is key is the interaction *between* the ants. And so it goes with the brain. A neuron is simply a specialized cell, just like other cells in your body, but with some specializations that allow it to grow processes and propagate electrical signals. Like an ant, an individual brain cell just runs its local program its whole life, carrying electrical signals along its membrane, spitting out neurotransmitters when the time comes for it, and being spat upon by the neurotransmission of other cells. That's it. It lives in darkness. Each neuron spends its life embedded in a network of other cells, simply responding to signals. It doesn't know if it's involved in moving your eyes to read Shakespeare, or moving your hands to play Beethoven. It doesn't know about you. Although your goals, intentions, and abilities are completely dependent on the existence of these little neurons, they live on a smaller scale, with no awareness of the thing they have come together to build.

But get enough of these basic brain cells together, interacting in the right ways, and the mind emerges.

Everywhere you look you can find systems with emergent properties. No single hunk of metal on an airplane has the property of flight, but when you arrange the pieces in the right way, flight emerges. Pieces and parts of a system can be individually quite simple. It's all about their interaction. In many cases, the parts themselves are replaceable.

WHAT IS REQUIRED FOR CONSCIOUSNESS?

Although the theoretical details are not yet worked out, the mind seems to emerge from the interaction of the billions of pieces and parts of the brain. This leads to a fundamental question: can a mind emerge from anything with lots of interacting parts? For example, could a city be conscious? After all, a city is built on the interactions between elements. Think of all the signals moving through a city: telephone wires, fiber-optic lines, sewers carrying waste, every handshake between humans, every traffic light, and so on. The scale of interaction in a city is on a par with the human brain. Of course, it would be very hard to know if a city were conscious. How could it tell us? How could we ask it?

To answer a question like this requires a deeper question: for a network to experience consciousness, does it need more than just a number of parts—but instead a very particular structure to the interactions?

Professor Giulio Tononi at the University of Wisconsin is working to answer exactly that question. He has proposed a quantitative definition of consciousness. It's not enough, he thinks, that there are pieces and parts interacting. Instead, there has to be a certain organization underlying this interaction.

To research consciousness in a laboratory setting, Tononi uses transcranial magnetic stimulation (TMS) to compare activity in the brain when it's awake and when it's in deep sleep (when, as we saw in Chapter 1, your consciousness disappears). By introducing a burst of electrical current into the cortex, he and his team can then track how the activity spreads.

When a subject is awake, and consciously aware, a complex pattern of neural activity spreads out from the focus of the TMS pulse. Long-lasting ripples of activity spread to different cortical areas, unmasking widespread connectivity across the network. In contrast, when the person is in deep sleep, the same TMS pulse stimulates only a very local area, and the activity dies down quickly. The network has lost much of its connectivity. This same result is seen when a person is in a coma: activity spreads very little, but as the person emerges over weeks into consciousness, the activity spreads more widely.

Tononi believes this is because when we are awake and conscious, there is widespread communication between different cortical areas; in contrast, the unconscious state of sleep is characterized by a lack of communication across areas. In his framework, Tononi suggests that a conscious system requires a perfect balance of enough complexity to represent very different states (this is called differentiation) and enough connectivity to have distant parts of the network be in tight communication with one another (called integration). In his framework, the balance of differentiation and integration can be quantified, and he proposes that only systems in the right range experience consciousness.

If his theory turns out to be correct, it would give a noninvasive assessment of the level of consciousness in coma patients. It may also give us the means to tell whether inanimate systems have consciousness. So the answer to the question of whether a city was conscious could be answered: it would depend on whether the information flow is arranged in the right way—with

CONSCIOUSNESS AND NEUROSCIENCE

Take a moment to think about private, subjective experience: the show that only happens inside someone's head. For example, when I bite a peach while watching a sunrise, you can't know the exact experience I'm having internally; you can only guess based on experiences you've had. My conscious experience is mine, and yours is yours. So how can it be studied using the scientific method?

In recent decades, researchers have set out to illuminate the "neural correlates" of consciousness—that is, the exact patterns of brain activity that are present every time a person is having a particular experience, and present only when they're having that experience.

Take the ambiguous picture of a duck/rabbit. Like the old woman/young woman figure in Chapter 4, its interesting property is that you can only experience one interpretation at a time, but not both at once. So in the moments that you're having the experience of a rabbit, what precisely is the signature of activity in your brain? When you switch to the duck, what is your brain doing differently? Nothing has changed on the page, so the only thing changing must be the details of brain activity that produce your conscious experience.

just the perfect amount of differentiation and integration.

Tononi's theory is compatible with the idea that human consciousness could escape its biological origins. In this view, although consciousness evolved along a particular path that resulted in a brain, it doesn't have to be built on top of organic matter. It could just as easily be made of silicon, assuming the interactions are organized in the right way.

UPLOADING CONSCIOUSNESS

If the software of the brain is the critical element to a mind—and not the details of the hardware—then, in theory, we could shift ourselves off the substrate of our bodies. With powerful enough computers simulating the interactions in our brains, we could upload. We could exist digitally by running ourselves as a simulation, escaping the biological wetware from which we've arisen, becoming nonbiological beings. That would be the single most significant leap in the history of our species, launching us into the era of transhumanism.

Imagine what it could look like to leave your body behind and enter a new existence in a simulated world. Your digital existence could look like any life you wanted. Programmers could make any virtual world for you—worlds in which you can fly, or live underwater, or feel the winds of a different planet. We could run our virtual brains as fast or slow as we wanted, so our minds could span immense swaths of time or turn seconds of computing time into billions of years of experience.

A technical hurdle to successful uploading is that the simulated brain must be able to modify itself. We would need not only the pieces and parts, but also the physics of their ongoing interactions—for example, the activity of transcription factors that travel to the nucleus and cause gene expression, the dynamic changes in location and strength of the synapses, and so on. Unless your simulated experiences changed the structure of your simulated brain, you would be unable to form new memories and would have no sense of the passage of time. Under those circumstances, would there be any point to immortality?

If uploading proves to be possible, it would open up the capacity to reach other solar systems. There are at least a hundred billion other galaxies in our cosmos, each of which contains a hundred billion stars. We've already spotted thousands of exoplanets orbiting those stars, some of which have conditions quite like our Earth. The difficulty lies in the impossibility that our current fleshy bodies will ever get to those exoplanets—there's simply no foreseeable way that we will be able to travel those kinds of distances in space and time. However, because you can pause a simulation, shoot it out into space, and reboot it a thousand years later when it arrives at a planet, it would seem to your consciousness that you were on Earth, you had a launch, and then instantly you found yourself on a new planet. Uploading would be equivalent to achieving the physics dream of finding a wormhole, allowing us to get from one part of the universe to another in a subjective instant.

ARE WE ALREADY LIVING IN A SIMULATION?

Maybe what you would choose for your simulation is something very much like your present life on Earth, and that simple thought has led several philosophers to wonder whether we're already living in a simulation. While that idea seems fantastical, we already know how easily we can be fooled into accepting our reality: every night we fall asleep and have bizarre dreams—and while we're there we believe those worlds entirely.

Questions about our reality are not new. Two thousand three hundred years ago, the Chinese philosopher Chuang Tzu dreamed he was a butterfly. Upon waking, he considered this question: how would I know if I was Chuang Tzu dreaming I'm a butterfly—or instead, if right now I'm a butterfly dreaming I'm a man named Chuang Tzu?

The French philosopher René Descartes wrestled with a different version of this same problem. He wondered how we could ever know if what we experience is the real reality. To make the problem clear, he entertained a thought experiment: How do I know I'm not a brain in a vat? Maybe someone is stimulating that brain in just the right way to make

UPLOADING: IS IT STILL YOU?

If biological algorithms are the important part of what makes us who we are, rather than the physical stuff, then it's a possibility that we will someday be able to copy our brains, upload them, and live forever in silica. But there's an important question here: is it really you? Not exactly. The uploaded copy has all your memories and believes it was you, just there, standing outside the computer, in your body. Here's the strange part: if you die and we turn on the simulation one second later, it would be a transfer. It would be no different from beaming up in *Star Trek*, when a person is disintegrated, and then a new version is reconstituted a moment later. Uploading may not be all that different from what happens to you each night when you go to sleep: you experience a little death of your consciousness, and the person who wakes up on your pillow the next morning inherits all your memories, and believes him or herself to be you.

me believe that I'm here and I'm touching the ground and seeing those people and hearing those sounds. Descartes concluded there might not be any way to know. But he also realized something else: there's some *me* at the center trying to figure all this out. Whether or not I'm a brain in a vat, I'm pondering the problem. I'm thinking about this, and therefore I exist.

INTO THE FUTURE

In the coming years we will discover more about the human brain than we can describe with our current theories and frameworks. At the moment we're surrounded by mysteries: many that we recognize and many we haven't yet registered. As a field, we have vast uncharted waters ahead of us. As always in science, the important thing is to run the experiments and assess the results. Mother Nature will then tell us which approaches are blind alleys, and which move us further down the road of understanding the blueprints of our own minds.

Only one thing is certain: our species is just at the beginning of something, and we don't fully know what it is. We're at an unprecedented moment in history, one in which brain science and technology are coevolving. What happens at this intersection is poised to change who we are.

For thousands of generations, humans have lived the same sort of life cycle over and over: we're born, we control a fragile body, we enjoy a small strip of sensory reality, and then we die.

Science may give us the tools to transcend that evolutionary story. We can now hack our own hardware, and as a result our brains don't need to remain as we've inherited them. We're capable of inhabiting new kinds of sensory realities and new kinds of bodies. Eventually we may even be able to shed our physical forms altogether.

Our species is just now discovering the tools to shape our own destiny.

Who we become is up to us.

ACKNOWLEDGMENTS

Just as the magic of the brain emerges from the interaction of many parts, so the book and television series of *The Brain* arose from collaboration among many people.

Jennifer Beamish was a pillar of the project, tirelessly managing people, juggling the evolving content of the television series in her head and managing the nuances of multiple personalities at once. Beamish was irreplaceable; this project simply wouldn't exist without her. The second pillar of this project was Justine Kershaw. The expertise and bravery with which Justine envisions big projects, runs a company (Blink Films), and manages scores of people is a continual inspiration to me. Throughout the filming of the television series, we had the pleasure of working with a team of absurdly talented directors: Toby Trackman, Nic Stacey, Julian Jones, Cat Gale, and Johanna Gibbon. I never cease to be astonished by how perceptive they are to shifting patterns of emotion, color, luminance, setting, and tone. Together, we had the pleasure of working with connoisseurs of the visual world, directors of photography Duane McClune, Andy Jackson, and

Mark Schwartzbard. The fuel for the series was provided daily by the resourceful and energetic assistant producers Alice Smith, Chris Baron, and Emma Pound.

For this book, I had the pleasure of working with Katy Follain and Jamie Byng at Canongate Books, consistently one of the bravest and most insightful publishing houses in the world. Equally, it's an honor and pleasure to work with my American editor Dan Frank at Pantheon Books; he acts in equal parts as my friend and adviser.

I have endless gratitude to my parents for their inspiration: my father is a psychiatrist, my mother a biology teacher, and they are both lovers of teaching and learning. They constantly stimulated and cheered my growth in the direction of a researcher and communicator. Although we almost never watched television in my childhood, they made sure to sit me down for Carl Sagan's *Cosmos*; this project has deep roots that stretch back to those evenings.

I thank the brilliant and industrious students and postdocs in my neuroscience laboratory for coping with my upside-down schedule during the filming of the show and the writing of the book.

Finally, and most importantly, I thank my beautiful wife Sarah for supporting me, cheering me on, putting up with me, and holding down the fort as I undertook this project. I'm a lucky man that she believes in the importance of this endeavor as much as I do.

ENDNOTES

Chapter 1—Who Am I?

The teenage brain and increased self-consciousness

Somerville, L. H., Jones, R. M., Ruberry, E. J., Dyke, J. P., Glover, G., and Casey, B. J. (2013) "The medial prefrontal cortex and the emergence of self-conscious emotion in adolescence." *Psychological Science*, 24(8), 1554–62.

Note the authors also found increased connection strength between the medial prefrontal cortex and another brain region called the striatum. The striatum, and its network of connections, is involved in turning motivations into actions. The authors suggest this connectivity may explain why social considerations strongly drive behavior in teens and why they're more likely to take risks in the presence of peers.

Bjork, J. M., Knutson, B., Fong, G. W., Caggiano, D. M., Bennett, S. M., and Hommer, D. W. (2004) "Incentive-elicited brain activation in adolescents: similarities and differences from young adults." *The Journal of Neuroscience*, 24(8), 1793–1802.

Spear, L. P. (2000) "The adolescent brain and age-related behavioral manifestations." *Neuroscience and Biobehavioral Reviews*, 24(4), 417–63.

Heatherton, T. F. (2011) "Neuroscience of self and self-regulation." *Annual Review of Psychology*, 62, 363–90.

Cab drivers and The Knowledge

Maguire, E. A., Gadian, D. G., Johnsrude, I. S., Good, C. D., Ashburner, J., Frackowiak, R. S., and Frith, C. D. (2000) "Navigation-related structural change in the hippocampi of taxi drivers." *Proceedings of the National Academy of Sciences of the United States of America*, 97(8), 4398–4403.

Number of cells in the brain

Also, note that there are an equal number of neurons and glial cells, about eighty-six billion of each in the whole human brain.

Azevedo, F. A. C., Carvalho, L. R. B., Grinberg, L. T., Farfel, J. M., Ferretti, R. E. L., Leite, R. E. P., and Herculano-Houzel, S. (2009) "Equal numbers of neuronal and nonneuronal cells make the human brain an isometrically scaled-up primate brain." *The Journal of Comparative Neurology*, 513(5), 532–41.

Estimates of the numbers of connections (synapses vary widely), but a quadrillion (that is, one thousand billion) is a reasonable ballpark estimate, if one assumes almost one hundred billion neurons with about ten thousand connections each. Some neuronal types have fewer synapses; others (such as Purkinje cells) have many more—about 200,000 synapses each.

Also see the encyclopedic collection of numbers on Eric Chudler's "Brain Facts and Figures": faculty.washington.edu/chudler/facts.html.

Musicians have better memory

Chan, A. S., Ho, Y. C., and Cheung, M. C. (1998) "Music training improves verbal memory." *Nature*, *396*(6707).

Jakobson, L. S., Lewycky, S. T., Kilgour, A. R., and Stoesz, B. M. (2008) "Memory for verbal and visual material in highly trained musicians." *Music Perception*, *26*(1), 41–55.

Einstein's brain and the Omega sign

Falk, D. (2009) "New information about Albert Einstein's Brain." *Frontiers in Evolutionary Neuroscience, 1.*

See also Bangert, M. and Schlaug, G. (2006) "Specialization of the specialized in features of external human brain morphology." *The European Journal of Neuroscience, 24*(6), 1832–4.

Memory of the future

Schacter, D. L., Addis, D. R., and Buckner, R. L. (2007) "Remembering the past to imagine the future: the prospective brain." *Nature Reviews Neuroscience, 8*(9), 657–61.

Corkin, S. (2013) *Permanent Present Tense: The Unforgettable Life Of The Amnesic Patient.* Basic Books.

Nun study

Wilson, R. S. et al. "Participation in cognitively stimulating activities and risk of incident Alzheimer disease." *Jama* 287.6 (2002), 742–48.

Bennett, D. A. et al. "Overview and findings from the religious orders study." *Current Alzheimer Research* 9.6 (2012), 628.

In their autopsy samples, the researchers found that half of the people with no cognitive troubles had signs of brain pathology, and one-third met the pathologic threshold for Alzheimer's disease. In other words, they found widespread signs of disease in the brains of the deceased—but these pathologies only accounted for about half of an individual's likelihood of cognitive decline. For more on the Religious Orders Study, see www.rush.edu/services-treatments/alzheimers-disease-center/religious-orders-study.

Mind–Body problem

Descartes, R. (2008) *Meditations on First Philosophy* (Michael Moriarty translation of 1641 ed.). Oxford University Press.

Chapter 2—What is Reality?

Visual illusions
Eagleman, D. M. (2001) "Visual illusions and neurobiology." *Nature Reviews Neuroscience*, 2(12), 920–6.

Prism goggles
Brewer, A. A., Barton, B., and Lin, L. (2012) "Functional plasticity in human parietal visual field map clusters: adapting to reversed visual input." *Journal of Vision*, 12(9), 1398.

Note that after the experiment has concluded and volunteers remove their goggles, it takes a day or two for them to return to normal proficiency as the brain refigures everything out.

Wiring the brain by interacting with the world
Held, R. and Hein, A. (1963) "Movement-produced stimulation in the development of visually guided behavior." *Journal of Comparative and Physiological Psychology*, 56 (5), 872–6.

Synchronizing the timing of signals
Eagleman, D. M. (2008) "Human time perception and its illusions." *Current Opinion in Neurobiology*, 18(2), 131–36.

Stetson C., Cui, X., Montague, P. R., and Eagleman, D. M. (2006) "Motor-sensory recalibration leads to an illusory reversal of action and sensation." *Neuron*, 51(5), 651–9.

Parsons, B., Novich S. D., and Eagleman, D. M. (2013) "Motor-sensory recalibration modulates perceived simultaneity of cross-modal events." *Frontiers in Psychology*, 4:46.

Hollow mask illusion
Gregory, Richard (1970) *The Intelligent Eye*. London: Weidenfeld & Nicolson.

Króliczak, G., Heard, P., Goodale, M. A., and Gregory, R. L. (2006) "Dissociation of perception and action unmasked by the hollow-face illusion." *Brain Res*, 1080 (1): 9–16.

As an interesting side note, people with schizophrenia are less susceptible to seeing the hollow mask illusion.

Keane, B. P., Silverstein, S. M., Wang, Y., and Papathomas, T. V. (2013) "Reduced depth inversion illusions in schizophrenia are state-specific and occur for multiple object types and viewing conditions." *J Abnorm Psychol*, 122 (2): 506–12.

Synesthesia

Cytowic, R. and Eagleman, D. M. (2009) *Wednesday is Indigo Blue: Discovering the Brain of Synesthesia.* Cambridge, MA: MIT Press.

Witthoft N., Winawer J., and Eagleman, D. M. (2015) "Prevalence of learned grapheme-color pairings in a large online sample of synesthetes." *PLoS ONE*, 10(3), e0118996.

Tomson, S. N., Narayan, M., Allen, G. I., and Eagleman, D. M. (2013) "Neural networks of colored sequence synesthesia." *Journal of Neuroscience*, 33(35), 14098–106.

Eagleman, D. M., Kagan, A. D., Nelson, S. N., Sagaram, D., and Sarma, A. K. (2007) "A standardized test battery for the study of Synesthesia." *Journal of Neuroscience Methods*, 159, 139–45.

Timewarp

Stetson, C., Fiesta, M., and Eagleman, D. M. (2007) "Does time really slow down during a frightening event?" *PloS One*, 2(12), e1295.

Chapter 3—Who's in control?

The power of the unconscious brain

Eagleman, D. M. (2011) *Incognito: The Secret Lives of the Brain.* Pantheon.

A handful of concepts I chose to include in The Brain *overlap with material in* Incognito. *This includes the cases of Mike May, Charles Whitman, and Ken Parks, as well as Yarbus's eye-tracking experiment, the trolley dilemma, the mortgage meltdown, and the Ulysses contract. In constructing the scaffolding for the present project, these contact points were deemed to be tolerable in part because the topics are discussed in a different manner and often for distinct purposes.*

Dilated eyes and attractiveness

Hess, E. H. (1975) "The role of pupil size in communication." *Scientific American*, 233(5), 110–12.

Flow state

Kotler, S. (2014) *The Rise of Superman: Decoding the Science of Ultimate Human Performance*. Houghton Mifflin Harcourt.

Subconscious influences on decision making

Lobel, T. (2014) *Sensation: The New Science of Physical Intelligence*. Simon & Schuster.

Williams, L. E. and Bargh, J. A. (2008) "Experiencing physical warmth promotes interpersonal warmth." *Science*, 322(5901), 606–7.

Pelham, B. W., Mirenberg, M. C., and Jones, J. T. (2002) "Why Susie sells seashells by the seashore: implicit egotism and major life decisions." *Journal of Personality and Social Psychology*, 82, 469–87.

Chapter 4—How do I decide?

Decision making

Montague, R. (2007) *Your Brain is (Almost) Perfect: How We Make Decisions*. Plume.

Coalitions of neurons

Crick, F. and Koch, C. (2003) "A framework for consciousness." *Nature Neuroscience*, 6(2), 119–26.

Trolley dilemma
Foot, P. (1967) "The problem of abortion and the doctrine of the double effect." *Reprinted in Virtues and Vices and Other Essays in Moral Philosophy (1978).* Blackwell.

Greene, J. D., Sommerville, R. B., Nystrom, L. E., Darley, J. M., and Cohen, J. D. (2001) "An fMRI investigation of emotional engagement in moral judgment." *Science*, 293(5537), 2105–8.

Note that emotions are measurable physical responses caused by things happening. Feelings, on the other hand, are the subjective experiences that sometimes accompany these bodily markers—what people commonly think of as the sensations of happiness, envy, sadness, and so on.

Dopamine and unexpected reward
Zaghloul, K. A., Blanco, J. A., Weidemann, C. T., McGill, K., Jaggi, J. L., Baltuch, G. H., and Kahana, M. J. (2009) "Human substantia nigra neurons encode unexpected financial rewards." *Science*, 323(5920), 1496–9.

Schultz, W., Dayan, P., and Montague, P. R. (1997) "A neural substrate of prediction and reward." *Science*, 275(5306), 1593–9.

Eagleman, D. M., Person, C., and Montague, P. R. (1998) "A computational role for dopamine delivery in human decision-making." *Journal of Cognitive Neuroscience*, 10(5), 623–30.

Rangel, A., Camerer, C., and Montague, P. R. (2008) "A framework for studying the neurobiology of value-based decision making." *Nature Reviews Neuroscience*, 9(7), 545–56.

Judges and parole decisions
Danziger, S., Levav, J., and Avnaim-Pesso, L. (2011) "Extraneous factors in judicial decisions." *Proceedings of the National Academy of Sciences of the United States of America*, 108(17), 6889–92.

Emotions in decision making
Damasio, A. (2008) *Descartes' Error: Emotion, Reason and the Human Brain.* Random House.

The power of now
Dixon, M. L. (2010) "Uncovering the neural basis of resisting immediate gratification while pursuing long-term goals." *The Journal of Neuroscience*, 30(18), 6178–9.

Kable, J. W., and Glimcher, P. W. (2007) "The neural correlates of subjective value during intertemporal choice." *Nature Neuroscience*, 10(12), 1625–33.

McClure, S. M., Laibson, D. I., Loewenstein, G., and Cohen, J. D. (2004) "Separate neural systems value immediate and delayed monetary rewards." *Science*, 306(5695), 503–7.

The power of the immediate applies not just to things right now, but also right here. Consider this scenario proposed by philosopher Peter Singer: as you're about to tuck into a sandwich, you look out the window and see a child on the sidewalk, starving, a tear running down his gaunt cheek. Might you give up your sandwich to the child, or would you simply eat it yourself? Most people feel happy to offer the sandwich. But right now, in Africa, there is that same child, starving, just like the boy on the corner. All it would take is a click of your mouse to send $5, the equivalent cost of that sandwich. Yet it's likely that you haven't sent over any sandwich money to him today, or even recently, despite your charitableness in the first scenario. Why haven't you acted to help him? It's because the first scenario puts the child right in front of you. The second requires you to imagine the child.

Willpower
Muraven, M., Tice, D. M., and Baumeister, R. F. (1998) "Self-control as a limited resource: regulatory depletion patterns." *Journal of Personality and Social Psychology*, 74(3), 774.

Baumeister, R. F. and Tierney, J. (2011) *Willpower: Rediscovering the Greatest Human Strength*. Penguin.

Politics and disgust

Ahn, W.-Y., Kishida, K. T., Gu, X., Lohrenz, T., Harvey, A., Alford, J. R., and Dayan, P. (2014) "Nonpolitical images evoke neural predictors of political ideology." *Current Biology*, 24(22), 2693–9.

Oxytocin

Scheele, D., Wille, A., Kendrick, K. M., Stoffel-Wagner, B., Becker, B., Güntürkün, O., and Hurlemann, R. (2013) "Oxytocin enhances brain reward system responses in men viewing the face of their female partner." *Proceedings of the National Academy of Sciences*, 110(50), 20308–313.

Zak, P. J. (2012) *The Moral Molecule: The Source of Love and Prosperity.* Random House.

Decisions and society

Levitt, S. D. (2004) "Understanding why crime fell in the 1990s: four factors that explain the decline and six that do not." *Journal of Economic Perspectives*, 163–90.

Eagleman, D. M. and Isgur, S. (2012). "Defining a neurocompatibility index for systems of law." In *Law of the Future*, Hague Institute for the Internationalisation of Law. 1(2012), 161–172.

Real-time feedback in neuroimaging

Eagleman, D. M. (2011) *Incognito: The Secret Lives of the Brain.* Pantheon.

Chapter 5—Do I need you?

Reading intention into others

Heider, F. and Simmel, M. (1944) "An experimental study of apparent behavior." *The American Journal of Psychology*, 243–59.

Empathy

Singer, T., Seymour, B., O'Doherty, J., Stephan, K., Dolan, R., and Frith, C. (2006) "Empathic neural responses are modulated by the perceived fairness of others." *Nature*, 439(7075), 466–9.

Singer, T., Seymour, B., O'Doherty, J., Kaube, H., Dolan, R., and Frith, C. (2004) "Empathy for pain involves the affective but not sensory components of pain." *Science*, 303(5661), 1157–62.

Empathy and outgroups

Vaughn, D. A. and Eagleman, D. M. (2010) "Religious labels modulate empathetic response to another's pain." Society for Neuroscience abstract.

Harris, L. T. and Fiske, S. T. (2011). "Perceiving humanity." In A. Todorov, S. Fiske, and D. Prentice (eds.). *Social Neuroscience: Towards Understanding the Underpinnings of the Social Mind.* Oxford Press.

Harris, L. T. and Fiske, S. T. (2007) "Social groups that elicit disgust are differentially processed in the mPFC." *Social Cognitive Affective Neuroscience*, 2, 45–51.

Circuitry of the brain devoted to other brains

Plitt, M., Savjani, R. R., and Eagleman, D. M. (2015) "Are corporations people too? The neural correlates of moral judgments about companies and individuals." *Social Neuroscience*, 10(2), 113–25.

Babies and trust

Hamlin, J. K., Wynn, K., and Bloom, P. (2007) "Social evaluation by preverbal infants." *Nature*, 450(7169), 557–59.

Hamlin, J. K., Wynn, K., Bloom, P., and Mahajan, N. (2011) "How infants and toddlers react to antisocial others." *Proceedings of the National Academy of Sciences*, 108(50), 19931–36.

Hamlin, J. K. and Wynn, K. (2011) "Young infants prefer prosocial to antisocial others." *Cognitive Development*, 26(1):30-39. doi:10.1016/j.cogdev.2010.09.001.

Bloom, P. (2013) *Just Babies: The Origins of Good and Evil.* Crown.

Reading emotion by simulating others' faces

Goldman, A. I. and Sripada, C. S. (2005) "Simulationist models of face-based emotion recognition." *Cognition*, 94(3).

Niedenthal, P. M., Mermillod, M., Maringer, M., and Hess, U. (2010) "The simulation of smiles (SIMS) model: embodied simulation and the meaning of facial expression." *The Behavioral and Brain Sciences*, 33(6), 417–33; discussion 433–80.

Zajonc, R. B., Adelmann, P. K., Murphy, S. T., and Niedenthal, P. M. (1987) "Convergence in the physical appearance of spouses." *Motivation and Emotion*, 11(4), 335–46.

Regarding the TMS experiment with John Robison, Professor Pascual-Leone reports: "We don't know exactly what happened neurobiologically, but I think it now offers the opportunity for us to understand what behavioral modifications, what interventions might be possible to learn from [John's case] that we can then teach others."

Botox diminishes the ability to read faces

Neal, D. T. and Chartrand, T. L. (2011) "Embodied emotion perception amplifying and dampening facial feedback modulates emotion perception accuracy." *Social Psychological and Personality Science*, 2(6), 673–8.

The effect is a small one, but significant: the Botox users showed 70 percent accuracy in identifying the emotions, while the control group averaged 77 percent.

Baron-Cohen, S., Wheelwright, S., Hill, J., Raste, Y., and Plumb, I. (2001) "The 'Reading the Mind in the Eyes' test revised version: A study with normal adults, and adults with Asperger syndrome or high-functioning autism." *Journal of Child Psychology and Psychiatry*, 42(2), 241–51.

Romanian orphans

Nelson, C. A. (2007) "A neurobiological perspective on early human deprivation." *Child Development Perspectives*, 1(1), 13–18.

The pain of social exclusion

Eisenberger, N. I., Lieberman, M. D., and Williams, K. D. (2003) "Does rejection hurt? An fMRI study of social exclusion." *Science*, 302(5643), 290–92.

Eisenberger, N. I. and Lieberman, M. D. (2004) "Why rejection hurts: a common neural alarm system for physical and social pain." *Trends in Cognitive Sciences*, 8(7), 294–300.

Solitary confinement

Beyond our interviews with Sarah Shourd for the television series, see also:

Pesta, A. (2014) "Like an Animal": Freed U.S. Hiker Recalls 410 Days in Iran Prison. NBC News.

Psychopaths and the prefrontal cortex

Koenigs, M. (2012) "The role of prefrontal cortex in psychopathy." *Reviews in the Neurosciences*, 23(3), 253–62.

The areas that activate differently in psychopaths are two neighboring regions of the midline part of the prefrontal cortex: the ventromedial PFC and anterior cingulate cortex. These areas are both commonly seen in studies of social and emotional decision making, and they're down-regulated in psychopathy.

Blue eyes brown eyes experiment
Transcript quoted from *A Class Divided*, original broadcast: March 26, 1985. Produced and directed by William Peters. Written by William Peters and Charlie Cobb.

Chapter 6—Who will we be?

Number of cells in the human body
Bianconi, E., Piovesan, A., Facchin, F., Beraudi, A., Casadei, R., Frabetti, F., and Canaider, S. (2013) "An estimation of the number of cells in the human body." *Annals of Human Biology*, 40(6), 463–71.

Brain plasticity
Eagleman, D. M. (in press). *LiveWired: How the Brain Rewires Itself on the Fly.* Canongate.

Eagleman, D. M. (March 17, 2015). David Eagleman: "Can we create new senses for humans?" TED conference. [Video file]. http://www.ted.com/talks/david_eagleman_can_we_create_new_senses_for_humans?

Novich, S. D. and Eagleman, D. M. (2015) "Using space and time to encode vibrotactile information: toward an estimate of the skin's achievable throughput." *Experimental Brain Research*, 1–12.

Cochlear implants
Chorost, M. (2005) *Rebuilt: How Becoming Part Computer Made Me More Human.* Houghton Mifflin Harcourt.

Sensory substitution
Bach-y-Rita, P., Collins, C., Saunders, F., White, B., and Scadden, L. (1969) "Vision substitution by tactile image projection." *Nature*, 221(5184), 963–4.

Danilov, Y., and Tyler, M. (2005) "Brainport: an alternative input to the brain." *Journal of Integrative Neuroscience*, 4(04), 537–50.

The connectome: making a map of all the connections in a brain

Seung, S. (2012) *Connectome: How the Brain's Wiring Makes Us Who We Are.* Houghton Mifflin Harcourt.

Kasthuri, N. et al. (2015) "Saturated reconstruction of a volume of neocortex." Cell: in press.

Image credit for volume of mouse brain: Daniel R. Berger, H. Sebastian Seung and Jeff W. Lichtman.

The Human Brain Project

The Blue Brain Project: http://bluebrain.epfl.ch. The Blue Brain team has joined with approximately eighty-seven international partners to drive the Human Brain Project (HBP).

Computation on other substrates

Building computational devices on strange substrates has a long history: an early analog computer called the Water Integrator was built in the Soviet Union in 1936.

More recent examples of water computers use microfluidics—see:

Katsikis, G., Cybulski, J. S., and Prakash, M. (2015) "Synchronous universal droplet logic and control." *Nature Physics.*

Chinese Room Argument

Searle, J. R. (1980) "Minds, brains, and programs." *Behavioral and Brain Sciences*, 3(03), 417–24.

Not everyone agrees with this interpretation of the Chinese Room. Some people suggest that although the operator doesn't understand Chinese, the system as a whole (the operator plus the books) does understand Chinese.

Leibniz's mill argument

Leibniz, G. W. (1989) *The Monadology*. Springer.

Here's the argument in Leibniz's own words:

> *Moreover, it must be confessed that perception and that which depends upon it are inexplicable on mechanical grounds, that is to say, by means of figures and motions. And supposing there were a machine, so constructed as to think, feel, and have perception, it might be conceived as increased in size, while keeping the same proportions, so that one might go into it as into a mill. That being so, we should, on examining its interior, find only parts which work one upon another, and never anything by which to explain a perception. Thus it is in a simple substance, and not in a compound or in a machine, that perception must be sought for. Further, nothing but this (namely, perceptions and their changes) can be found in a simple substance. It is also in this alone that all the internal activities of simple substances can consist.*

Ants

Hölldobler, B. and Wilson, E. O. (2010) *The Leafcutter Ants: Civilization by Instinct*. WW Norton & Company.

Consciousness

Tononi, G. (2012) *Phi: A Voyage from the Brain to the Soul*. Pantheon Books.

Koch, C. (2004) *The Quest for Consciousness*. New York.

Crick, F. and Koch, C. (2003) "A framework for consciousness." *Nature Neuroscience*, 6(2), 119–26.

GLOSSARY

Action Potential A brief (one millisecond) event in which the voltage across a neuron reaches a threshold, causing a propagating chain reaction of ion exchange across the cell membrane. Eventually this causes neurotransmitter release at the terminals of the axon. Also known as a spike.

Alien Hand Syndrome A disorder resulting from a treatment for epilepsy known as a corpus callosotomy, in which the corpus callosum is cut, disconnecting the two cerebral hemispheres of the brain, also known as split-brain surgery. This disorder causes unilateral and sometimes intricate hand movements without the patient feeling they have volitional control of the movements.

Axon The anatomical output projection of a neuron capable of conducting electrical signals from the cell body.

Cerebrum Human brain areas including the large, undulate exterior cerebral cortex, hippocampus, basal ganglia, and olfactory bulb. Development of this area in higher-order mammals contributes to more advanced cognition and behavior.

Cerebellum A smaller anatomical structure that sits below the cerebral cortex at the rear of the head. This area of the brain is essential for fluid motor control, balance, posture, and possibly some cognitive functions.

Computational Hypothesis of Brain Function A framework proposing that the interactions in the brain are implementing computations, and that the same computations, if run on a different substrate, would equally give rise to the mind.

Connectome A three-dimensional map of all neuronal connections in the brain.

Corpus Callosum A strip of nerve fibers located in the longitudinal fissure between the two cerebral hemispheres that enables communication between them.

Dendrites The anatomical input projections of a neuron that carry electrical signals initiated by neurotransmitter release from other neurons to the cell body.

Dopamine A neurotransmitter in the brain linked to motor control, addiction, and reward.

Electroencephalography (EEG) A technique used to measure electrical activity at millisecond resolution in the brain by connecting conductive electrodes to the scalp. Each electrode captures the summation of millions of neurons underlying the electrode. This method is used to capture fast changes in brain activity in the cortex.

Functional Magnetic Resonance Imaging (fMRI) A neuroimaging technique that detects brain activity with second resolution by measuring blood flow in the brain with millimeter resolution.

Galvanic Skin Response A technique that measures changes in the autonomic nervous system which occur when someone experiences something new, stressful, or intense even if below conscious awareness. In practice, a machine is hooked up to the fingertip and the electrical properties of the skin are monitored that change along with activity in the skin sweat glands.

Glial Cell Specialized cells in the brain that protect neurons by providing nutrients and oxygen to them, removing waste, and generally supporting them.

Neural Of or relating to the nervous system or neurons.

Neuron A specialized cell found in both the central and peripheral nervous systems, including brain, spinal cord, and sensory cells, that communicates to other cells using electrochemical signals.

Neurotransmitter Chemicals that are released from one neuron to another recipient neuron, usually across a synapse. These are found in the central and peripheral nervous systems including the brain, spinal cord, and sensory neurons throughout the body. Neurons may release more than one neurotransmitter.

Parkinson's Disease A progressive disorder characterized by movement difficulties and tremors that is caused by the deterioration of dopamine-producing cells in a midbrain structure called the substantia nigra.

Plasticity The brain's ability to adapt by creating new or modifying existing neural connections. The capability of the brain to exhibit plasticity is important after an injury in order to compensate for any acquired deficits.

Sensory Substitution An approach to compensate for an impaired sense in which sensory information is fed into the brain through unusual sensory channels. For example, visual information is converted into vibrations on the tongue or auditory information is converted into patterns of vibrations on the torso, allowing an individual to see or hear respectively.

Sensory Transduction Signals from the environment, such as photons (sight), air compression waves (hearing) or scent molecules (smell) are converted (transduced) into action potentials by specialized cells. It is the first step by which information from outside the body is received by the brain.

Split-brain Surgery Also known as a corpus callosotomy, the corpus callosum is severed as a measure to control epilepsy not cured by other means. This surgery removes the communication between the two cerebral hemispheres.

Synapse The space typically between an axon of one neuron and a dendrite of another neuron where communication between neurons occurs by release of neurotransmitters. Axon–axon and dendrite–dendrite synapses also exist.

Transcranial Magnetic Stimulation (TMS) A noninvasive technique used to stimulate or inhibit brain activity using a magnetic pulse to induce small electric currents in underlying neural tissue. This technique is typically used to understand the influence of brain areas in neural circuits.

Ulysses Contract An unbreakable contract used to bind oneself to a potential future goal made when one understands one may not have the ability to make a rational choice at that time.

Ventral Tegmental Area A structure comprised of mostly dopaminergic neurons located in the midbrain. This area plays a critical role in the reward system.